Advisory Editor: Keith Stead, M.A., D.Phil.
University of Exeter

Polymers: Chemistry and Physics
of Modern Materials

J.M.G. COWIE
Professor of Chemistry
University of Stirling

Intertext Books

Published by
International Textbook Company Limited
24 Market Square, Aylesbury, Bucks HP20 1TL

First published 1973

ISBN 0 7002 0222 6

Printed in Great Britain
by Billing & Son Ltd, Worcester

Polymers: Chemistry and Physics of Modern Materials

Contents

Preface

It has been said that a science graduate entering a science-based industry has a better than 30 per cent chance of being involved with work relating to polymers in one form or another. If this is a fact then a basic knowledge of the subject would be generally useful and it is regrettable that a lack of uniformity prevails in the level of polymer science taught to undergraduates in the various universities and polytechnics. This is partly a result of the original acceptance of polymer science as a predominantly postgraduate course which has retarded its general acceptance as part of the undergraduate curriculum. While there are many excellent texts written for the postgraduate on all specialised aspects of the subject, fewer books exist which are designed specifically for undergraduates and of these the majority tend to concentrate on either the chemical or physical aspects. It is my firm belief that a broadly based understanding of the science is a most useful acquisition for every science student and that an undergraduate text should cover the subject area as widely as possible.

The inter-disciplinary nature of polymer science is obvious. Polymers are ultimately materials with characteristic mechanical and physical properties which are controlled by the structure and the methods of synthesis. Consequently a scientist or engineer gains most from the subject if the inter-disciplinary approach is emphasised from the beginning, but of course there must be a starting point. Bearing that in mind this book is developed in the sequence: preparation, characterization, physical and mechanical properties, and culminates in a coverage of structure-property relations. Unfortunately certain limitations have to be imposed if a reasonably priced text is to be provided for undergraduate use and not the least of these is length. Such space limitations lead to sins of omission and these are inevitably personal. No mention is made of the extensive technology involved or of polymer stability and degradation. Also there is a deliberate lack of in-depth treatment throughout, which should be sought, when required, in the specialist text. Nevertheless it is hoped that this book will provide a reasonably broad coverage at the undergraduate level and overlap to some extent with postgraduate courses, thereby providing a foundation on which a graduate may subsequently build.

My thanks are due to both Dr. Keith Stead and Dr. W.V. Steele for their constructive comments and criticisms which have helped to improve the manuscript. Finally I would like to dedicate this book to my long-suffering, but patient family—Ann, Graeme and Christian.

J.M.G. Cowie

Stirling 1973

CHAPTER 1

Introduction

1.1 Historical development

Ever since the late 19th century, when the click of the first celluloid billiard balls
heralded a reprieve for hosts of elephants, to the present more modern sound of
disposable plastic articles rumbling down the rubbish dumps of the world, the
growth of polymer science and the polymer industry has been inexorable.

The development of this branch of science has not been uniform, however,
but rather more exponential, with the most dramatic expansion occurring in the
last twenty years. Before examining the scientific content of these years it is
instructive to trace the historical progression from the first uncertain steps of
"manufacture without scientific understanding" to the modern, relatively
sophisticated "molecular engineering" which attempts to make a material
according to predetermined specifications.

The roots of both the science and the industry can be acknowledged to have
been established early in the 19th century in England when, in 1820, Hancock
discovered the effect of masticating natural rubber. He followed this in 1843
with a patent for the sulphur crosslinking process which reduced the undesirable
tackiness of rubber and improved its elastic properties. This patent was actually
superseded, independently, by Goodyear in the U.S.A. in 1839, who later, in
1851, discovered ebonite thereby laying a foundation stone in the development
of thermosetting plastics.

Most of the earlier work was carried out, not unexpectedly, on naturally
occurring polymers. At the Great International Exhibition in London in 1862
Alexander Parkes displayed articles moulded from a mixture of cellulose nitrate
and castor oil, which earned the award of a medal for quality. This may have
encouraged him to establish a small company to manufacture these items in
larger quantities, but the problem of scaling up his successful bench process
to a factory operation proved too great and the company collapsed two years

later. The difficulties of scaling up are of course a constant source of aggravation and are epitomized in the resigned tones of the verse:

> "First the test-tube, then the pail,
> Then the semi-working scale,
> Ever bigger, ever faster,
> Faster, faster, then — disaster!"

Parkes can be credited then, without reproach, with the first attempt to produce plastics commercially. The problem was solved in 1870 by John W. Hyatt in the U.S.A., who used camphor instead of castor oil as the plasticizer to soften the cellulose nitrate and produced celluloid. This work was actually stimulated by his experiments on billiard balls covered with collodion. He was searching for a better material after receiving complaints that the violent cannoning of the balls occasionally led to a mild explosion. One billiard saloon owner in the West commented that while this did not spoil the game unduly the sudden noise rather upset his patrons, who reacted each time by quickly drawing their guns.

While celluloid rapidly became a commercial success and held the market almost exclusively for about thirty years, it is highly inflammable and more stable materials were soon being sought. Cellulose acetate and casein based plastics were then developed, followed soon after by bakelite in 1910. This arose from patents filed by the Belgian chemist Leo Baekeland who capitalized on previous work on phenol-formaldehyde resins, by discovering how to control the reaction and fabricate the product. Bakelite was an immediate success and encouraged further research in this area which led to the development of the urea-formaldehyde resins.

In 1917 the shortages of raw materials, brought about by blockading during World War I, forced German chemists to develop a synthetic "methyl" rubber from dimethyl butadiene. The product was a miserable substitute for the natural product but served to launch what is now a thriving synthetic rubber industry.

Until then real progress had been hampered by the lack of any fundamental knowledge concerning the structure of these materials. Although large molecular weights had been reported for the natural products, rubber and starch, and cellulose nitrate, these were regarded with scepticism by most scientists. Up to about 1930 the more generally accepted view was that these apparently large molecules were colloidal aggregates of smaller molecules held in a micellar structure by secondary forces. The basic units in both cellulose and rubber were known, but cellulose was thought to be a cyclic tetrasaccharide and rubber merely a ring composed of two isoprene molecules.

Eventually, as a result of the vigorous pioneering work of Hermann Staudinger, chemists gradually began to accept his revolutionary concept that these molecules were actually long sequences of smaller structural units held together by covalent bonds to form large chains or macromolecules. For his monumental work in establishing polymer science Staudinger received the Nobel prize in 1953, by which time the foundations of the modern science and industry had been firmly laid by himself, Mark, Carothers, Flory, Meyer, and many others. From 1930 onwards the progress gained momentum. Polyethylene was discovered accidentally around 1932 by ICI workers, R. O. Gibson and J. Swallow. In 1934 W. H. Carothers, working for DuPont, made nylon and subsequently produced a superb series of publications on condensation polymerization reactions. By the late 1930s Hill and Crawford of ICI had developed poly(methyl methacrylate) — "perspex" — and both polystyrene and poly(vinyl chloride) were in commercial production. After the second World War an acceleration in both research and industrial output began and it continues unabated so that in 1970 the annual production of synthetic polymers in Britain alone was in excess of a million tons while the world total can be estimated at about ten times that figure.

Manfred Gordon has aptly called polymer science a "revolution in chains", and happily it is a revolution which is still active and regularly revitalized by new advances. It has contributed to the expansion of man's experience of his environment, extended his comforts and, sadly, increased some of his problems — but basically it is exciting science and we can now begin to explore some of its fundamental aspects.

1.2 Classification
Because of the diversity of function and structure found in the field of macromolecules, it is advantageous to draw up some scheme which groups the materials under convenient headings, and one way of doing this is shown below.

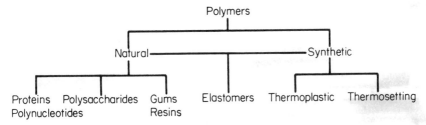

Natural polymers usually have more complex structures than synthetic polymers and we shall deal almost exclusively with the latter group. Elastomers can be either natural or man-made and are classified here as a common sub-group. The more general term elastomer is used to describe rubberlike materials, because there now exists a wide variety of synthetic products, whose structures differ markedly from the naturally occurring rubber, but whose elastic properties are comparable to, and sometimes better than, the original.

1.3 Some basic definitions
In order to place polymer science in the proper perspective we must examine the subject on as broad a basis as possible. It is useful to consider polymers first on the molecular level then as materials. These considerations can be interrelated

by examining the various aspects in the sequence: synthesis, characterization, mechanical behaviour, and application, but before discussing the detailed chemistry and physics some of the fundamental concepts must be introduced to provide essential background to such a development. We need to know what a polymer is and how it is named and prepared. It is also useful to identify which physical properties are important and so it is necessary to define the molar mass, obtain an appreciation of the molecular size and shape, and recognise the important transition temperatures.

A *polymer* is a large molecule constructed from many smaller structural units called *monomers,* covalently bonded together in any conceivable pattern. In certain cases it is more accurate to call the structural or repeat unit a *monomer residue* because atoms are eliminated from the simple monomeric unit during some polymerization processes.

The essential requirement for a small molecule to qualify as a monomer or "building block" is the possession of two or more bonding sites, through which each can be linked to other monomers to form the polymer chain. The number of bonding sites is referred to as the *functionality*. Monomers such as a hydroxy-acid $(HO-R-COOH)$ or vinyl chloride $(CH_2=CHCl)$ are bifunctional. The hydroxyacid will condense with the other hydroxyacid molecules through the $-OH$ and $-COOH$ groups to form a linear polymer, and the polymerization reaction in this case consists of a series of simple organic reactions similar to

$$ROH + R'COOH \rightleftharpoons R'COOR + H_2O$$

The double bond of the vinyl compound is also bifunctional as activation by a free radical or an ion leads to polymer formation

$$CH_2=CHCl + R^\cdot \rightarrow RCH_2-CHCl-CH_2-CHCl \rightsquigarrow$$

Bifunctional monomers form linear macromolecules but if the monomers are polyfunctional, *i.e.* have three or more bonding sites as in glycerol CH_2OH. $CHOH. CH_2OH$, branched macromolecules can be produced. These may even develop into large three-dimensional networks containing both branches and crosslinks.

When only one species of monomer is used to build a macromolecule the product is called a *homopolymer,* normally referred to simply as a polymer. If the chains are composed of two types of monomer unit, the material is known as a *copolymer,* and if three different monomers are incorporated in one chain a *terpolymer* results.

Copolymers prepared from bifunctional monomers can be subdivided further into four main categories:

(i) Random copolymers where the distribution of the two monomers in the chain is essentially random.

$$\rightsquigarrow AAABABBABABBBBABAAB\rightsquigarrow$$

(ii) Alternating copolymers with a regular placement along the chain.

$$\sim\!\!\sim\!ABABABABAB\!\sim\!\!\sim$$

(iii) Block copolymers comprised of substantial sequences or blocks of each.

$$\sim\!\!\sim\!AAAAAABBBBBAAAA\!\sim\!\!\sim$$

(iv) Graft copolymers in which blocks of one monomer are grafted on to a backbone of the other as branches.

```
        B                     B
        B                     B
        B                     B
        B                     B
     AAAAAAAAAAAAAAAAAAA
              B
              B
              B
```

1.4 Synthesis of polymers

A process used to convert monomer molecules into a polymer is called a *polymerization* and the two most important groups are step-growth and addition. A step-growth polymerization is used for monomers with functional groups such as $-OH, -COOH, -COCl$, *etc.* and is normally, but not always, a succession of condensation reactions. Consequently the majority of polymers formed in this way differ slightly from the original monomers because a small molecule is eliminated in the reaction, *e.g.* the reaction between ethylene glycol and terephthalic acid produces a polyester better known as terylene

$$n\,HO(CH_2)_2OH + n\,HOOC-\!\!\left\langle\bigcirc\right\rangle\!-COOH \rightarrow \left(-O(CH_2)_2O\,.\,C-\!\left\langle\bigcirc\right\rangle\!-C-\right)_n$$
$$+ (2n-1)\,H_2O$$

The addition polymerizations, for olefinic monomers, are chain reactions which convert the monomers into polymers by stimulating the opening of the double bond with a free radical or ionic initiator. The product then has the same chemical composition as the starting material, *e.g.* acrylonitrile produces polyacrylonitrile without the elimination of a small molecule.

$$n\,CH_2\!\!=\!\!CHCN \rightarrow \sim\!(CH_2CHCN)_{\overline{n}}\sim$$

The length of the molecular chains, which will depend on the reaction conditions, can be obtained from measurements of molar masses.

1.5 Nomenclature

The least ambiguous method of naming a polymer is based on its source. However, a wide variety of trade names are commonly used. The prefix poly is attached to the name of the monomer in addition polymers, and so polyethylene, poly-acrylonitrile, polystyrene denote polymers prepared from these single monomers. When the monomer has a multi-worded name or has a substituted parent name then this is enclosed in brackets and prefixed with poly, *e.g.* poly(methylmetha-crylate), poly (vinyl chloride), poly (ethylene oxide), *etc.*

Polymers prepared by self-condensation of a single monomer such as ω-amino lauric acid, are named in a similar manner, but this polymer, poly(ω-amino lauric acid), (sometimes known as nylon-12) can also be prepared by a ring-opening reaction using lauryl lactam and could then be called poly(lauryl lactam). Both names are correct.

Many condensation polymers are synthesized from two monomers which form a repeating residue. Thus ethylene glycol and terephthalic acid form a polyester

$$\left[OCH_2CH_2OCO-\!\!\!\left\langle\!\bigcirc\!\right\rangle\!\!-\!CO \right]_n$$

where the acid is considered to be the parent compound and the name poly(ethylene terephthalate) is given to the polymer. The polyurethane formed from ethylene glycol and phenylene diisocyanate

$$\left[(CH_2)_2OCONH-\!\!\!\left\langle\!\bigcirc\!\right\rangle\!\!-\!NHCOO \right]_n$$

is called poly(ethylene phenylene urethane), where the structural groups attached to the parent compound, or class of compound, are enclosed in the brackets.

It has been proposed that these and other condensation and step-growth polymers, prepared from two monomers, should be treated as copolymers and named instead as poly(ethylene glycol-*co*-terephthalic acid) and poly(ethylene glycol-*co*-phenylene diisocyanate). While this would remove any ambiguity, this convention is at present reserved for linear copolymers prepared by addition techniques, *e.g.* poly(acrylonitrile-*co*-vinyl acetate). For the various types of copolymers, that is random, alternating, block, and graft, the abbreviations *-co-, -alt-, -b-,* and *-g-* are used.

1.6 Average molar masses and distributions*

One of the most important features which distinguishes a synthetic high polymer from a simple molecule is the inability to assign an exact molar mass to a polymer. This is a consequence of the fact that in a polymerization reaction the length of the chain formed is determined entirely by random events. In

---

*The quantity molar mass is used throughout this text instead of the dimensionless quantity molecular weight which is usual in polymer chemistry. All the equations in later sections evaluate molar mass rather than the dimensionless quantity molecular weight.

a condensation reaction, it depends on the availability of a suitable reactive group and in an addition reaction, on the lifetime of the chain carrier. Inevitably, because of the random nature of the growth process, the product is a mixture of chains of differing length – a *distribution* of chain lengths – which in many cases can be calculated statistically.

The polymer is characterized best by a molar mass distribution and the associated molar mass averages, rather than by a single molar mass. The typical distributions, shown in figure 1.1, can be described by a variety of averages. As the methods used for estimating molar mass of polymers employ different averaging procedures, it is safer to use more than one technique to obtain two or more averages and by doing so characterize the sample more fully.

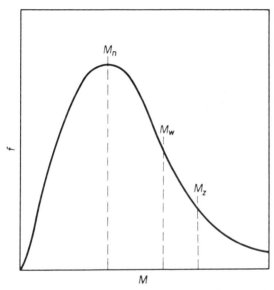

FIGURE 1.1. Typical distribution of molar masses for a synthetic polymer sample, where f is the fraction of polymer in each interval of M considered.

A colligative method, such as osmotic pressure, effectively counts the number of molecules present and provides a *number average* molar mass $<M>_n$ defined by

$$<M>_n = \frac{\Sigma N_i M_i}{\Sigma N_i} = \frac{\Sigma w_i}{\Sigma(w_i/M_i)} \qquad (1.1)$$

where N_i is the number of molecules of species i of molar mass M_i. The brackets $<>$ indicate that it is an average value, by convention these are normally omitted.

The alternative expression is in terms of the mass $w_i = N_i M_i / N_A$ if required, where N_A is Avogadro's constant.

From light scattering measurements, a method depending on the size rather than the number of molecules, a weight average molar mass $\langle M \rangle_w$ is obtained. This is defined as

$$\langle M \rangle_w = \frac{\Sigma N_i M_i^2}{\Sigma N_i M_i} = \frac{\Sigma w_i M_i}{\Sigma w_i} \qquad (1.2)$$

Statistically $\langle M \rangle_n$ is simply the first moment, and $\langle M \rangle_w$ is the ratio of the second to the first moment, of the number distribution.

A higher average, the z-average given by

$$\langle M \rangle_z = \frac{\Sigma N_i M_i^3}{\Sigma N_i M_i^2} = \frac{\Sigma w_i M_i^2}{\Sigma w_i M_i} \ , \qquad (1.3)$$

can be measured in the ultracentrifuge which also yields another useful average, the $(z + 1)$-average,

$$\langle M \rangle_{z+1} = \frac{\Sigma N_i M_i^4}{\Sigma N_i M_i^3} \ , \qquad (1.4)$$

often required when describing mechanical properties.

A numerical example serves to highlight the differences in the various averages. Consider a hypothetical polymer sample composed of chains of four distinct molar masses, 100 000, 200 000, 500 000, and 1 000 000 g mol^{-1} in the ratio 1:5:3:1, then

$$M_n/\text{g mol}^{-1} = \frac{(1 \times 10^5) + (5 \times 2 \times 10^5) + (3 \times 5 \times 10^5) + (1 \times 10^6)}{1 + 5 + 3 + 1} = 3{\cdot}6 \times 10^5$$

$$M_w/\text{g mol}^{-1} = \frac{\{1 \times (10^5)^2\} + \{5 \times (2 \times 10^5)^2\} + \{3 \times (5 \times 10^5)^2\} + \{1 \times (10^6)^2\}}{(1 \times 10^5) + (5 \times 2 \times 10^5) + (3 \times 5 \times 10^5) + (1 \times 10^6)} = 5.45 \times 10^5$$

and $M_z = 7.22 \times 10^5$ g mol^{-1}

The breadth of the distribution can often be gauged by establishing the *heterogeneity index* (M_w/M_n). For many polymerizations the most probable value is about 2.0, but both larger and smaller values can be obtained and it is at best only a rough guide.

An alternative method of describing the chain length of a polymer is to measure the *average degree of polymerization x*. This represents the number of monomer units or residues in the chain and is given by

$$x = M/M_0, \qquad (1.5)$$

where M_0 is the molar mass of monomer or residue and M is the appropriate average molar mass. Hence the x average depends on which average is used for M. (To avoid confusion between the mole fraction x and the average degree of polymerization x, the latter will always be subscripted as x_n or x_w to indicate the particular M used in equation (1.5).)

1.7 Size and shape

Some measure of the polymer size is obtained from the molar mass, but what is the actual length of a chain and what shape does it adopt? We can begin to answer these questions by first considering a simple molecule such as butane and examining the behaviour when the molecule is rotated about the bond joining carbon 2 to carbon 3.

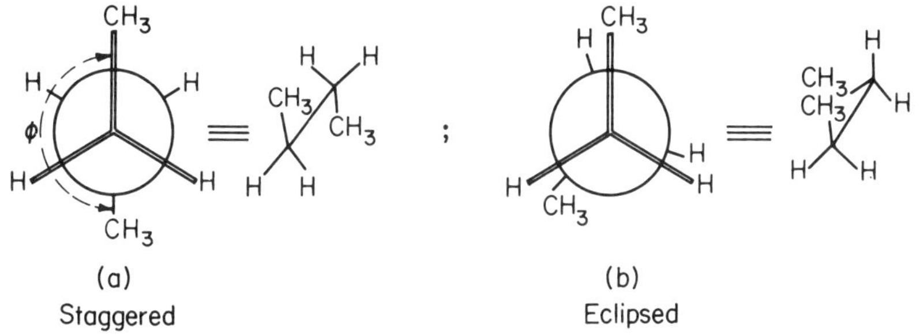

(a)
Staggered

(b)
Eclipsed

FIGURE 1.2. Newman and "saw horse" projections for *n*-butane, (a) a staggered state with $\phi = \pi$ and (b) an eclipsed position.

The Newman and "saw horse" projections show the *trans* position in figure 1.2a with the "dihedral angle" $\phi = 180°$. This is the most stable conformation with the greatest separation between the two methyl groups. Rotation about the C_2—C_3 bond alters ϕ and moves the methyl groups past the opposing hydrogen atoms so that an extra repulsive force is experienced when an eclipsed position (figure 1.2b) is reached.

The progress of rotation can be followed by plotting the change in potential energy $V(\phi)$ as a function of the dihedral angle, as shown in figure 1.3. The resulting diagram for butane exhibits three minima at $\phi = \pi$, $\pi/3$, and $5\pi/3$ called the *trans* and $\pm$ *gauche* states respectively, and the greater depth of the *trans* position indicates that this is the position of maximum stability. Although the *gauche* states are slightly less stable, all three minima can be regarded as discrete rotational states. The maxima correspond to the eclipsed positions and are angles of maximum instability. These diagrams will vary with the type of molecule and need not be symmetrical, but the butane diagram is very similar to that for the simple polymer polyethylene $-(CH_2-CH_2-)_n$, if the $-CH_3$ groups are replaced by the two sections of the chain adjoining the bond of rotation. The backbone of this polymer is composed of a chain of tetrahedral carbon atoms covalently bonded to each other so that the molecule can be represented as an extended all *trans*

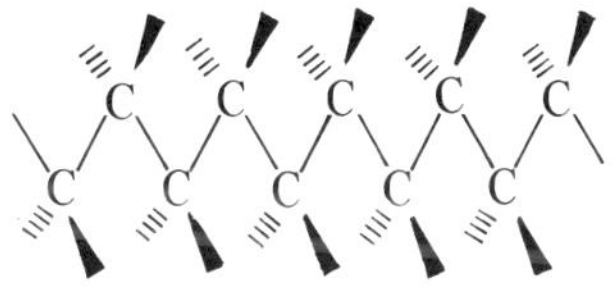

zig-zag chain. For a typical value of $M = 1.6 \times 10^5$ g mol^{-1}, the chain contains 10 000 carbon atoms; thus in the extended zig-zag state, assuming a tetrahedral angle of 109° and a bond length of 0.154 nm, the chain would be about 1260 nm long and 0.3 nm diameter. Magnified a million times, the

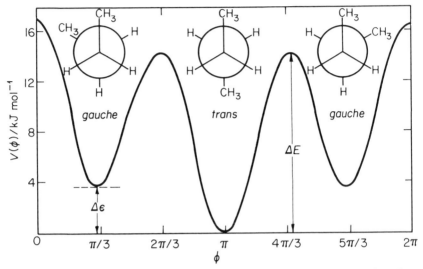

FIGURE 1.3. Potential energy $V(\phi)$ as a function of the dihedral angle ϕ for *n*-butane.

chain could be represented by a piece of wire 126 × 0.03 cm. This means that polyethylene is a long threadlike molecule, but how realistic is the extended all *trans* conformation? As every group of four atoms in the chain has a choice of three possible stable rotational states, a total of $3^{10\,000}$ shapes are available to

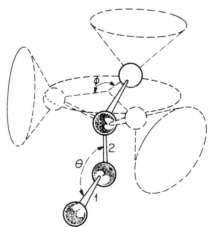

FIGURE 1.4. Diagrammatic representation of the cones of revolution available to the third and fourth bonds of a simple carbon chain with a fixed bond angle θ.

this particular chain, only one of which is the all *trans* state. So, in spite of the fact that the all *trans* extended conformation has the lowest energy, the most probable conformation will be some kind of randomly coiled state – assuming that no external ordering forces are present and that the rotation about the carbon bonds is in no way impeded. The many possible coiled forms are generated simply by allowing the chain to rotate into a *gauche* position which moves the atom out of the plane of the adjacent bonds. This is shown more clearly (see figure 1.4) by considering the various cones of revolution available to a chain over only two bonds. The distribution of *trans* (*t*) and *gauche* (*g*) states along a chain will be a function of the temperature and the relative stability of these states. Consequently there is an unequal distribution of each. The ratio of the number of *trans* n_t to *gauche* n_g states is then governed by a Boltzmann factor and

$$n_g/n_t = 2 \exp(-\Delta\epsilon/kT), \qquad (1.6)$$

where k is the Boltzmann constant, $\Delta\epsilon$ is the energy difference between the two minima, and the 2 arises because of the $\pm$ *gauche* states available. For polyethylene $\Delta\epsilon$ is about 3.34 kJ mol^{-1}, and values of (n_g/n_t) for 100, 200, and 300 K are 0.036, 0.264, and 0.524 respectively, showing that the chain becomes less extended and more coiled as the temperature increases. Because of the possibility of rotation about the carbon bonds, the chain is in a state of perpetual motion, constantly changing shape from one coiled conformation to another form, equally probable at the given temperature. The speed of this wriggling varies with temperature (and from one polymer to another) and dictates many of the physical characteristics of the polymer, as we shall see later.

The height of the potential energy barrier ΔE determines the rate of bond interchange between the t and the g states and for polyethylene is about 16.7 kJ mol^{-1}. When ΔE is very high (about 80 kJ mol^{-1}), rotation becomes very difficult, but as the temperature is raised the fraction of molecules which possess energy in excess of ΔE increases and rotation from one state to another becomes easier.

Realistically then, a polymer chain is better represented by a loosely coiled ball (figure 1.5) than an extended rod. For the magnified polyethylene chain considered earlier a ball of about 4 cm diameter is a likely size.

The term *conformation* has been used here when referring to the three-dimensional geometric arrangement of the polymer, which changes easily when the bonds are rotated.

There is a tendency to use the term *configuration* in a synonymous sense, but as far as possible this will be reserved for the description of chains where the geometric variations can only be interchanged by breaking a bond.

1.8 The glass transition temperature T_g and the melting temperature T_m

At sufficiently low temperatures all polymers are hard rigid solids. As the temperature rises, each polymer eventually obtains sufficient thermal energy to enable its chains to move freely enough for it to behave like a viscous liquid (assuming no degradation has occurred).

There are two ways in which a polymer can pass from the solid to the liquid phase, depending on the internal organization of the chains in the sample. The

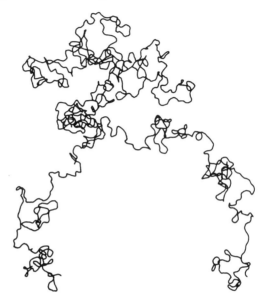

FIGURE 1.5. Random arrangement of a polyethylene chain containing 1000 freely rotating C—C bonds, in which each successive bond has been given a random choice of six equally spaced angular positions. (Treloar (1958), *Physics of Rubber Elasticity.*)

different types of thermal response, illustrated by following the change in specific volume, are shown schematically in figure 1.6.

A polymer may be completely amorphous in the solid state, which means that the chains in the specimen are arranged in a totally random fashion. The volume change in amorphous polymers follows the curve A—D. In the region C—D the polymer is a glass, but as the sample is heated it passes through a temperature T_g, called the *glass transition temperature,* beyond which it softens and becomes rubberlike. This is an important temperature because it marks the point where important property changes take place, *i.e.* the material may be more easily deformed or become ductile above T_g. A continuing increase in temperature along C—B—A leads to a change of the rubbery polymer to a viscous liquid.

In a perfectly crystalline polymer, all the chains would be incorporated in regions of three-dimensional order, called crystallites, and no glass transition would be observed, because of the absence of disordered chains in the sample. The crystalline polymer, on heating, would follow curve H—B—A; at T_m^o, melting would be observed and the polymer would become a viscous liquid.

Perfectly crystalline polymers are not encountered in practice and instead polymers may contain varying proportions of ordered and disordered regions in the sample. These semi-crystalline polymers usually exhibit both T_g and T_m, corresponding to the ordered and disordered portions and follow curves similar to F–E–G–A. As T_m^o is the melting temperature of a perfectly crystalline polymer of high molar mass, T_m is lower and more often represents a melting range, because the semi-crystalline polymer contains a spectrum of chain lengths and crystallites of various sizes with many defects. These imperfections act to depress the melting temperature and experimental values of T_m can depend on the previous thermal history of the sample.

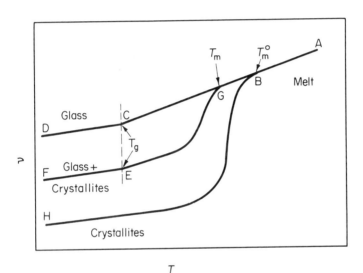

FIGURE 1.6. Schematic representation of the change of specific volume v of a polymer with temperature T for (i) a completely amorphous sample (A–C–D), (ii) a semi-crystalline sample (AGF), and (iii) a perfectly crystalline material (A–B–H).

Nevertheless, both T_g and T_m are important parameters, which serve to characterize a given polymer.

1.9 Elastomers, fibres, and plastics

A large number of synthetic polymers now exist covering a wide range of properties. These can be grouped into the three major classes, plastics, fibres, and elastomers, but there is no firm dividing line between the groups. However, some classification is useful from a technological viewpoint and one method of defining a member of these categories is to examine a typical stress-strain plot. Rigid plastics and fibres are resistant to deformation and are characterized by a high modulus and low percentage elongations. Elastomers readily undergo deformation and exhibit large reversible elongations under small applied stresses, *i.e.* they exhibit elasticity. The flexible plastics are intermediate in behaviour. An outline of the structure-property relations will be presented later, but before proceeding further with the more detailed science, we can profitably familiarize ourselves with some of the more common polymers and their uses.

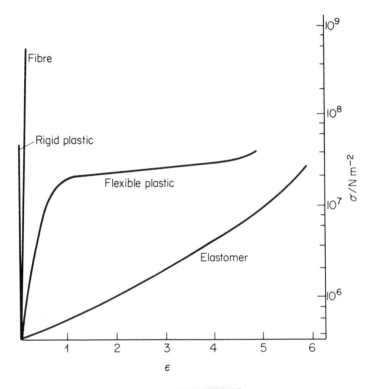

FIGURE 1.7. Typical stress–strain ($\sigma - \epsilon$) plots for a rigid plastic, a fibre, a flexible plastic, and an elastomer.

Some of these are presented in table 1.1 where an attempt is made to show that the lines of demarkation which are used to divide polymers into the three major groups are not clear cut.

A polymer normally used as a fibre may make a perfectly good plastic if no attempt is made to draw it into a filament. Similarly, a plastic, if used at a temperature above its glass transition and suitably crosslinked, may make a perfectly acceptable elastomer. In the following brief account of some of the more common plastics, fibres, and elastomers, the classification is based essentially on their major technological application under standard working conditions.

1.10 Fibre-forming polymers

While there are many fibre-forming polymers only a limited number have achieved great technological and commercial success. It is significant that these are polymers of long standing, and it has been suggested that further fibre

TABLE 1.1. Some common plastics, elastomers, and fibres

Elastomers	Plastics	Fibres
polyisoprene	polyethylene	
polyisobutylene	polytetrafluoroethylene	
polybutadiene	polystyrene	
	poly(methylmethacrylate)	
	phenol-formaldehyde	
	urea-formaldehyde	
	melamine-formaldehyde	
←————— poly(vinyl chloride)—————→		
←————— polyurethanes —————→		
←————— polysiloxanes —————→		
	←————— polyamide —————→	
	←————— polyester —————→	
	←————— polypropylene —————→	

research may involve the somewhat prosaic task of attempting to improve, modify, or reduce the cost of existing fibres, rather than to look for new and better alternatives. The commercially important fibres are listed in table 1.2; all are thermoplastic polymers.

The polyamides are an important group of polymers which include the naturally occurring proteins in addition to the synthetic nylons. The term nylon, originally a trade name, has now become a generic term of the synthetic polyamides, and the numerals which follow, *e.g.* nylon-6,6 , distinguish each polymer by designating the number of carbon atoms lying between successive amide groups in the chain. Thus nylon-6,10 is prepared from two monomers and has the structure

$$\left[\text{NH(CH}_2)_6\text{NHCO(CH}_2)_8\text{CO} \right]_n$$

with alternative sequences of six and ten carbon atoms between the nitrogen atoms, while nylon-6 is prepared from one monomer and has the repeat formula $\left[\text{NH(CH}_2)_5\text{CO} \right]_n$ with regular sequences of six carbon atoms between the nitrogen atoms. A nylon with two numbers is termed *dyadic* indicating that it contains both dibasic acid (or acid chloride) and diamine moieties, where the first number represents the diamine and the second the diacid used in the synthesis. The *monadic* nylons have one number, indicating that synthesis involved only one type of monomer. This terminology means that a poly (α-amino acid) would be nylon-2.

TABLE 1.2. Chemical structure of synthetic fibres

Polymer	Repeat unit	Trade names
Step-growth		
POLYAMIDES (Nylons) (Uses: drip-dry fabrics, cordage, braiding, bristles, and surgical sutures.)		
polycaprolactam	$+NH(CH_2)_5CO+_n$	Nylon-6, Perlon
poly(decamethylene carboxamide)	$+NH(CH_2)_{10}CO+_n$	Nylon-11, Rilsan
poly(hexamethylene adipamide)	$+NH(CH_2)_6NHCO(CH_2)_4CO+_n$	Nylon-6,6 , Bri-nylon
poly(m-phenylene isophthalamide)	$\left[NH{-}\bigcirc{-}NHCO{-}\bigcirc{-}CO \right]_n$	Nomex
POLYESTERS (Uses: fabrics, tyre-cord yarns, and yacht sails.)		
poly(ethylene terephthalate)	$\left[OC{-}\bigcirc{-}COO(CH_2)_2O \right]_n$	Terylene, Dacron
poly(cyclohexane 1,4-dimethylene terephthalate)	$\left[OCH_2{-}\bigcirc{-}CH_2OOC{-}\bigcirc{-}CO \right]_n$	Kodel

TABLE 1.2. Chemical structure of synthetic fibres

Polymer	Repeat Unit	Trade Names
POLYUREAS		
poly(nonamethylene urea)	$+NHCONH(CH_2)_9+_n$	Urylon
Addition		
ACRYLICS (Uses: fabrics and carpeting.)		
polyacrylonitrile	$+CH_2CHCN+_n$ (often as copolymer with $>$ 85 per cent acrylonitrile)	Orlon, Courtelle, Acrilan, Creslan
acrylonitrile copolymers	35 per cent $<$ acrylonitrile $<$ 85 per cent + vinyl chloride + vinylidene chloride	Dynel Verel
HYDROCARBONS (Uses: carpets and upholstery.)		
polyethylene	$+CH_2CH_2+_n$	Courlene, Vestolen
polypropylene (isotactic)	$+CH_2-CH-CH_3)_n$	Ulstron, Herculon, Meraklon
HALOGEN SUBSTITUTED OLEFINES (Uses: knitwear and protective clothing.)		
poly(vinyl chloride)	$+CH_2CHCl+_n$	Rhovyl, Valren
poly(vinylidene chloride)	$+CH_2CCl_2+_n$	Saran, Tygan
polytetrafluoroethylene	$+CF_2CF_2+_n$	Teflon, Polifen
VINAL (Uses: fibres, adhesives, paint, sponges, films, and plasma extender.)		
poly(vinyl alcohol)	$+CH_2CHOH+_n$ (normally crosslinked)	Vinylon, Kuralon, Mewlon

Terylene is the most important polyester. It exhibits high resilience, durability, and low moisture absorption, properties which contribute to its desirable "wash and wear" characteristics. The harsh feel of the fibre, caused by the stiffness of the chain, is overcome by blending it with wool and cotton.

The acrylics and modacrylics are the most important of the amorphous fibres. They are based on the acrylonitrile unit ─ CH$_2$CH(CN) ─ and are usually manufactured as copolymers. When the acrylonitrile content is 85 per cent or higher, the polymer is an *acrylic* fibre, but if this drops to between 35 and 85 per cent it is known as a *modacrylic* fibre. Vinyl chloride and vinylidene chloride are the most important comonomers and the copolymers produce high bulk yarns which can be subjected to a controlled shrinking process after fabrication. Once shrunk the fibres are dimensionally stable.

Silk-like qualities have always been sought after by the synthetic fibre chemist. The new cycloaliphatic polyamide "Qiana", with the probable structure

$$\left[\text{NH}\!-\!\bigcirc\!-\!\text{CH}_2\!-\!\bigcirc\!-\!\text{NH}\!-\!\text{CO}\!-\!(\text{CH}_2)_{\overline{m}}\ \text{CO}\right]_n$$

is said to have the aesthetic appeal of natural silk, and a silk-like fibre called "Chinon" has been prepared from a polyacrylonitrile-protein graft copolymer. This has been prepared by grafting acrylonitrile on to caseine and has many of the properties of natural silk.

1.11 Plastics

A plastic is rather inadequately defined as an organic high polymer capable of changing its shape on the application of a force and retaining this shape on removal of this force, *i.e.* a material in which a stress produces a non-reversible strain.

The main criterion is that plastic materials can be formed into complex shapes, often by the application of heat or pressure and a further sub-division into those which are *thermosetting* and those which are *thermoplastics* is useful. The thermosetting materials become permanently hard when heated above a critical temperature and will not soften again on reheating. They are usually crosslinked in this state. A thermoplastic polymer will soften when heated above T_g. It can then be shaped and on cooling will harden in this form. However, on reheating it will soften again and can be reshaped if required before hardening when the temperature drops. This cycle can be carried out repeatedly.

A number of the important thermoplastics are shown in table 1.3 together with a few examples of their more important uses, determined by the outstanding properties of each. Thus polypropylene, poly(phenylene oxide) and TPX have good thermal stability and can be used for items requiring sterilization. The optical qualities of polystyrene and poly(methyl methacrylate) are used in situations where transparency is a premium, while the low frictional coefficient and superb chemical resistance of poly tetrafluoroethylene make it useful in non-stick cookware and protective clothing. Low density polyethylene, while mechanically inferior to the high density polymer, has better impact resistance and can be used when greater flexibility is required, whereas the popularity of poly(vinyl chloride) lies in its unmatched ability to form a stable, dry, flexible material when plasticized. The polyamides and terylene are also important thermoplastics.

1.12 Thermosetting polymers

The thermoset plastics generally have superior abrasion and dimensional stability characteristics compared with the thermoplastics which have better flexural and impact properties. In contrast to the thermoplastics, thermosetting polymers, as the name implies, are changed irreversibly from fusible, soluble products into highly intractable crosslinked resins which cannot be moulded by flow and so must be fabricated during the crosslinking process. Typical examples are:

Phenolic resins are prepared by reacting phenols with aldehydes. They are used for electrical fitments, radio and television cabinets, heat resistant knobs for cooking utensils, game parts, buckles, handles, and a wide variety of similar items.

Amino resins are related polymers formed from formaldehyde and either urea or melamine. In addition to many of the uses listed above, they can be used to manufacture lightweight tableware and counter and table surfaces. Being transparent they can be filled and coloured using light pastel shades, whereas the phenolics are already rather dark and consequently have a more restricted colour range.

Thermosetting *polyester resins* are used in paints and surface coatings where oxidation during drying forms a crosslinked film which provides a tough resistant finish.

Epoxy resins are polyethers prepared from glycols and dihalides and find extensive use as surface coatings, adhesives, and flexible enamel-like finishes because of their combined properties of toughness, chemical resistance, and flexibility.

1.13 Elastomers

The modern elastomer industry was founded on the naturally occurring product isolated from the latex of the tree *Hevea brasiliensis*. It was first used by South American Indians and was called caoutchouc from the Indian name, but later, simply rubber, when it was discovered, by Priestley, that the material rubbed out pencil marks.

From the early 20th century, chemists have been attempting to synthesize materials whose properties duplicate or at least simulate those of natural rubber, and this has led to the production of a wide variety of synthetic elastomers.

TABLE 1.3. Thermoplastics.

Polymer	Repeat unit	Density (g cm^{-3})	Uses
Polyethylene (High Density) (Low Density)	—(CH$_2$CH$_2$)—	0.94 to 0.96 0.92	Household products, insulators, pipes, toys, bottles
Polypropylene	—(CH$_2$CH(CH$_3$))—	0.90	Waterpipes, integral hinges, sterilizable hospital equipment
Poly(4-methylpentene-1) (TPX)	—(CH$_2$CH)— $\mid$ CH$_2$ $\mid$ CH(CH$_3$)(CH$_3$)	0.83	Hospital and laboratory ware
Poly tetrafluoroethylene (PTFE)	—(CF$_2$CF$_2$)—	2.20	Non-stick surfaces, insulation, gaskets
Poly(vinyl chloride) (PVC)	—(CH$_2$CHCl)—	1.35 to 1.45	Records, bottles, house siding and eaves
Polystyrene	—(CH$_2$CH(C$_6$H$_5$))—	1.04 to 1.06	Lighting panels, lenses, wall tiles, flower pots
Poly(methyl methacrylate) (PMMA)	—(CH$_2$—C(CH$_3$)(COOCH$_3$))—	1.17 to 1.20	Bathroom fixtures, knobs, combs, illuminated signs
Polycarbonates	—(R.O.COO)—	1.20	Cooling fans, marine propellors, safety helmets
Poly(phenylene oxide)		1.06	Hot water fittings, sterilizable, medical, and surgical equipment

Some of these have become technologically important and are listed in table 1.4 together with their general uses.

Although a large number of synthetic elastomers are now available, natural rubber must still be regarded as the standard elastomer because of the excellently balanced combination of desirable qualities. Presently it accounts for almost 36 per cent of the total world consumption of elastomers and its gradual replacement by synthetic varieties is partly a result of demand outstripping natural supply.

The most important synthetic elastomer is styrene-butadiene (SBR) which accounts for 41 per cent of the world market in elastomers. It is used predominantly for vehicle tyres when reinforced with carbon black. Nitrile rubber (NBR) is a random copolymer of acrylonitrile (mass fraction 0.2 to 0.4) and butadiene and it is used when an elastomer is required which is resistant to swelling in organic solvents. The range of properties can be extended when styrene is also incorporated in the chain, forming ABS rubber. Butyl rubber (IIR) is prepared by copolymerizing small quantities of isoprene (3 parts) with isobutylene (97 parts). The elastic properties are poor but it is resistant to corrosive fluids and has a low permeability to gases. Polychloroprene possesses the desirable qualities of being a fire retardant and resistance to weathering, chemicals, and oils.

Elastomers which fail to crystallize on stretching must be strengthened by the addition of fillers such as carbon black. SBR, poly(ethylene-co-propylene) and the silicone elastomers fall in this category. While polyethylene is normally highly crystalline, copolymerization with propylene destroys the ordered structure and if carried out in the presence of a small quantity of non-conjugated diene (*e.g.* dicyclopentadiene) a crosslinking site is introduced. The material is an

amorphous random terpolymer which when crosslinked forms an elastomer with a high resistance to oxidation. Unfortunately it is incompatible with other elastomers and is unsuitable for blending.

The silicone elastomers have a low cohesive energy between the chains which results in poor thermoplastic properties and an unimpressive mechanical response. Consequently they are used predominantly in situations requiring temperature stability over a range of 190 to 570 K when conditions are unsuitable for other elastomers.

More recently, extensive use has been made of room temperature vulcanizing silicone rubbers. These are based on linear polydimethyl siloxane chains, with M ranging from 10^4 to 10^5 g mol^{-1}, and hydroxyl terminal groups. Curing can be achieved in a number of ways, either by adding a crosslinking agent and a metallic salt catalyst, such as tri- or tetra-alkoxysilane with stannous octoate, or by

TABLE 1.4. Some common elastomers and their uses

Polymer	Formula	Uses
Natural rubber (polyisoprene-*cis*)	$+CH_2-\underset{\underset{CH_3}{\mid}}{C}=CH-CH_2\xrightarrow{}_n$	General purposes
Polybutadiene	$-(CH_2-CH=CH-CH_2\xrightarrow{}_n$	Tyre treads
Butyl	$\left(-CH_2-\underset{\underset{CH_3}{\mid}}{\overset{\overset{CH_3}{\mid}}{C}}-\right)_n$	Inner tubes, cable sheathing, roofing, tank liners
SBR	$\left(CH_2-CH=CH-CH_2CH_2-\underset{C_6H_5}{CH}\right)_n$	Tyres, general purposes
ABS	$+CH_2-\underset{\underset{CN}{\mid}}{CH}-CH_2-\underset{CH-CH=CH-CH_2)_m}{CH}-C_6H_5)_n$	Oil hoses, gaskets, flexible fuel tanks
Polychloroprene	$\left(CH_2-\underset{\underset{Cl}{\mid}}{C}=CH-CH_2\right)_n$	Used when oil resistance, good weathering, and inflammability characteristics are needed
Silicones	$\left(-O-\underset{\underset{R}{\mid}}{\overset{\overset{R}{\mid}}{Si}}-\right)_n$	Gaskets, door seals, medical applications, flexible moulds
Polyurethanes	$+R_1-NHCO\cdot O\cdot R_2OOCHN\xrightarrow{}_n$	Printing rollers, sealing and jointing
EPR	$\sim(CH_2-CH_2)_m-\left(CH_2-\underset{\underset{CH_3}{\mid}}{CH}\right)_n\sim$	Window strips and channeling

incorporating in the mixture a crosslinking agent sensitive to atmospheric water which initiates vulcanization. The products are good sealing, encapsulating, and caulking materials; they make good flexible moulds and are excellent insulators. They have found a wide application it the building, aviation, and electronics industries.

Having briefly introduced the diversity of structure and property encountered in the synthetic polymers, we can now examine more closely the fundamental chemistry and physics of these materials.

General Reading

T. Alfrey and E. F. Gurnee, *Organic Polymers.* Prentice-Hall (1967).
F. W. Billmeyer, *Textbook of Polymer Science.* John Wiley and Sons (1962).
L. W. Chubb, *Plastics, Rubbers and Fibres.* Pan (1967).
E. W. Duck, *Plastics and Rubbers.* Butterworths (1971).
M. Gordon, *High Polymers.* Iliffe Books Ltd. (1963).
C. T. Greenwood and W. Banks, *Synthetic High Polymers.* Oliver and Boyd, (1968).
M. Kaufman, *Giant Molecules.'* Aldus (1968).
E. M. McCafferey, *Laboratory Preparation for Macromolecular Chemistry.* McGraw-Hill (1970).
W. R. Moore, *Introduction to Polymer Chemistry.* University of London Press (1963).
R. B. Seymour, *Introduction to Polymer Chemistry.* McGraw-Hill, (1971).
L. R. G. Treloar, *Introduction to Polymer Science.* Wykeham Publications (1970).

CHAPTER 2

Step-growth Polymerization

The classical subdivision of polymers into two main groups was made around 1929 by W. H. Carothers, who proposed that a distinction be made between polymers prepared by the stepwise reaction of monomers and those formed by chain reactions. These he called:

(1) *Condensation polymers,* characteristically formed by reactions involving the elimination of a small molecule, such as water, at each step; and

(2) *Addition polymers,* where no such loss occurred.

While these definitions were perfectly adequate at the time, it soon became obvious that notable exceptions existed and that a fundamentally sounder classification should be based on a description of the chain growth mechanism. It is preferable to replace the term condensation with step-growth or step-reaction. Reclassification as *step-growth* polymerization, now logically includes polymers such as polyurethanes, which grow by a step reaction mechanism without elimination of a small molecule.

In this chapter we shall examine the main features of step-growth polymerization, beginning with the simpler reactions which produce linear chains exclusively. This type of polymerization is used to produce some of the industrially important fibres such as the nylons and terylene. A brief discussion of the more complex branching reactions follows to illustrate how the thermosetting plastics are formed.

2.1 General reactions

In any reaction resulting in the formation of a chain or network of high molar mass, the functionality (see section 1. 3) of the monomer is of prime importance. In step-growth polymerization, a linear chain of monomer residues is obtained by the stepwise intermolecular condensation or addition of the reactive groups in bifunctional monomers. These reactions are analogous to simple reactions

involving monofunctional units as typified by a polyesterification reaction
involving a diol and a dibasic acid,

$$HO—R—OH + HOOC—R'—COOH = HO—R—OCO—R'—COOH + H_2O$$

If the water is removed as it is formed, no equilibrium is established and the first
stage in the reaction is the formation of a dimer which is also bifunctional. As the
reaction proceeds, longer chains, trimers, tetramers, and so on, will form through
other esterification reactions, all essentially identical in rate and mechanism,
until ultimately the reaction contains a mixture of polymer chains of large molar
masses M. However, the formation of samples with significantly large values of
M is subject to a number of rather stringent conditions which will be examined in
greater detail later in this chapter.

Two major groups, both distinguished by the type of monomer involved, can
be identified in step-growth polymerization. In the first group two polyfunctional
monomers take part in the reaction, and each possesses only one distinct type of
functional group as in the esterification reaction, or more generally:

$$A–A + B–B → (A–AB–B)$$

The second group is encountered when the monomer contains more than one
type of functional group such as a hydroxyacid (HO—R—COOH), represented
generally as $A –B$ where the reaction is

$$nA–B → (AB)_n$$

or $$n(HO—R—COOH) → H(ORCO)_{\overline{n}} OH.$$

A large number of step growth polymers have the basic structure

$$—\square—R—\square—R—\square—R—$$

where R can be $(CH_2)_x$ or ⬡ and the link $—\square—$ is one of three
important groups:

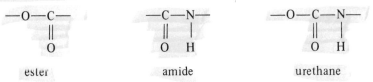

| ester | amide | urethane |

Other links and groups are involved in these reactions and some typical step-
reaction polymers are shown in table 2.1.

2.2 Reactivity of functional groups

One basic simplifying assumption proposed by Flory, when analysing the
kinetics of step-growth systems, was that all functional groups can be considered

TABLE 2.1. Typical step-growth polymerization reactions

Polymer	Reaction
Polyester	$n\mathrm{HO(CH_2)_x COOH} \rightarrow \mathrm{HO{+}(CH_2)_x{-}\underset{\underset{\displaystyle O}{\|}}{C}{-}O{\rightarrow}_n H} + (n-1)\mathrm{H_2O}$
Polyamide	$n\mathrm{NH_2{-}R{-}COOH} \rightarrow \mathrm{H{+}NH{-}R{-}CO{\rightarrow}_n OH} + (n-1)\mathrm{H_2O}$
	$n\mathrm{NH_2{-}R{-}NH_2} + n\mathrm{HOOC{-}R'{-}COOH} \rightarrow \mathrm{H{+}NH{-}R{-}NHCO{-}R'{-}CO{\rightarrow}OH} + (2n-1)\mathrm{H_2O}$
Polyurethanes	$n\mathrm{HO{-}R{-}OH} + \mathrm{OCN{-}R'{-}NCO} \rightarrow \mathrm{{+}OR{\cdot}O{\cdot}CONH{-}R'{-}NH{\cdot}CO{\rightarrow}_n}$
Polyanhydride	$n\mathrm{HOOC{-}R{-}COOH} \rightarrow \mathrm{HO{+}OC{-}R{-}CO{\cdot}O{\rightarrow}_n H} + (n-1)\mathrm{H_2O}$
Polysiloxane	$n\mathrm{HO{-}\underset{\underset{\displaystyle CH_3}{\|}}{\overset{\overset{\displaystyle CH_3}{\|}}{Si}}{-}OH} \rightarrow \mathrm{HO{-}\left[\underset{\underset{\displaystyle CH_3}{\|}}{\overset{\overset{\displaystyle CH_3}{\|}}{Si}}{-}O\right]_n H} + (n-1)\mathrm{H_2O}$
Phenol-formaldehyde	$n\,\mathrm{C_6H_4OH} + n\mathrm{CH_2O} \rightarrow$ structure $+ (n-1)\mathrm{H_2O}$

as being equally reactive. This implies that a monomer will react with both monomer or polymer species with equal ease.

The progress of the reaction can be illustrated in figure 2.1 where after 25 per cent reaction the number average chain length x_n is still less than two because monomers, being the most predominant species, will tend to react most often, and the reaction is mainly the formation of dimers and trimers. Even after 87.5 per cent of the reaction x_n will only be about eight and it becomes increasingly obvious that if long chains are required, the reaction must be pushed towards completion.

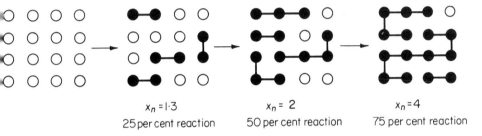

$x_n = 1.3$ $x_n = 2$ $x_n = 4$

25 per cent reaction   50 per cent reaction   75 per cent reaction

FIGURE 2.1. Diagrammatic representation of a step-growth polymerization.

2.3 Carothers equation

W. H. Carothers, the pioneer of step-growth reactions, proposed a simple equation relating x_n to a quantity p describing the extent of the reaction for linear polycondensations or polyadditions.

If N_0 is the original number of molecules present in an A–B monomer system and N the number of all molecules remaining after time t, then the total number of functional groups of either A or B which have reacted is $(N_0 - N)$. At that time t the extent of reaction p is given by

$$p = (N_0 - N)/N_0 \quad \text{or} \quad N = N_0(1 - p). \tag{2.1}$$

If we remember that $x_n = N_0/N$, a combination of expressions gives the *Carothers equation*,

$$x_n = 1/(1 - p). \tag{2.2}$$

This equation is also valid for an A–A + B–B reaction when one considers that in this case there are initially $2N_0$ molecules.

The Carothers equation is particularly enlightening when we examine the numerical relation between x_n and p; thus for $p = 0.95$ (*i.e.* 95 per cent conversion), $x_n = 50$ and when $p = 0.99$, then $x_n = 100$. In practical terms, it has been found that for a fibre forming polymer such as nylon-6,6 $+NH(CH_2)_6NHCO$ $(CH_2)_4CO+_n$ the value of M_n has to be about 12000 to 13 000 g mol^{-1} if a high tenacity fibre is to be spun, and as this corresponds to x_n = 53 to 58, the polymerization has to proceed beyond 95 per cent completion. Similarly for polyesters derived from ω-hydroxydecanoic acid, $H+O(CH_2)_9CO+_nOH$, x_n of about 150 is optimum for good fibres and so p must exceed 0.99.

2.4 Control of the molar mass

Quite obviously the control of the molar mass of the product of these reactions is very important. Very high molar mass material may be too difficult to process, while low molar mass polymer may not exhibit the properties desired in the end product, and one must be able to stop the reaction at the required value of p. Consequently the reactions are particularly demanding with respect to the purity of the reagents and accurate control of the amount of each species in the mixture is cardinal. It is symptomatic of these critical requirements that only four types of reaction usefully produce linear polymers with $M_n > 25000$ g mol^{-1}.

(1) *Schotten-Baumann reaction.* This involves the use of an acid chloride in an esterification or amidation; for example the so-called "nylon rope trick" reaction is an interfacial condensation between sebacoyl chloride and hexa-methylenediamine, producing a polyamide known as nylon-6,10.

$$nClCO(CH_2)_8\ COCl + nH_2N(CH_2)_6NH_2 \rightarrow +CO(CH_2)_8CONH(CH_2)_6NH+_n+(2n-1)H$$

The bifunctional acyl chloride is dissolved in CCl_4 and placed in a beaker. An aqueous alkaline solution of the bifunctional amine is layered on top and the nylon-6,10 which forms immediately at the interface can be drawn off as a continuous filament until the reagents are exhausted. The reaction has an S_N2 mechanism with the release of a proton to the base in the aqueous phase; the condensation is a successive stepwise reaction by this S_N2 route.

(2) *Salt dehydration.* Direct esterification requires high purity materials in equimolar amounts because esterifications rarely go beyond 98 per cent completion in practice. To overcome this, hexamethylene diamine and a dibasic acid such as adipic acid can be reacted to produce a nylon salt, hexamethylene diammonium adipate. A solution of 0.5 mol diamine in a mixture of 95 per cent ethanol (160 cm^3) and distilled water (60 cm^3) is added to 0.5 mol diacid dissolved in 600 cm^3 of 95 per cent ethanol over a period of 15 min. The mixture is stirred for 30 min during which time the nylon salt precipitates as a white crystalline solid. This can be recrystallized and should melt at 456 K. The pure salt can be converted into a polyamide by heating it under vacuum in a sealed tube, protected by wire gauze, at about 540 K in the presence of a small quantity of the diacid, e.g. 10 g salt to 0.55 g adipic acid is a suitable mixture. If a lower molar mass is desired, a monofunctional acid can replace the adipic acid and act as a chain terminator.

(3) *Ester interchange.* An alternative reaction is a trans-esterification in the presence of a proton donating or weak base catalyst such as sodium methoxide, *e.g.*

$$CH_3O-\underset{O}{\overset{\parallel}{C}}-\langle\bigcirc\rangle-\underset{O}{\overset{\parallel}{C}}-O-CH_3 + HO(CH_2)_2OH \rightarrow$$

$$\left[\begin{matrix}\underset{O}{\overset{\parallel}{C}}-\langle\bigcirc\rangle-\underset{O}{\overset{\parallel}{C}}-O-(CH_2)_2-O\end{matrix}\right]_n + CH_3OH$$

In this way ethylene glycol and dimethyl terephthalate produce terylene. Ester interchange is the most practical approach to polyester formation because of the faster reaction rate and use of more easily purified products. The formation of poly(ethylene terephthalate) is essentially a two-stage process. The first stage, at 380 to 470 K, is the formation of dimers and trimers, each with two hydroxyl end-groups, and during this formation the methanol is being distilled off. To complete the reaction the temperature is raised to 530 K and condensation of these oligomers produces a polymer with large M_n. The major advantage is that the stoichiometry is self-adjusting in the second stage.

(4) *Step polyaddition (Urethane formation).* Polyurethanes with large M_n can be prepared using a method based on the Wurtz alcohol test. In the presence of a basic catalyst, such as a diamine, ionic addition takes place, for example between 1,4-butanediol and 1,6-hexanediisocyanate:

$$HO(CH_2)_4OH + OCN(CH_2)_6 NCO \rightarrow \left[O(CH_2)_4-O-\underset{O}{\overset{\parallel}{C}}-\underset{H}{\overset{|}{N}}-(CH_2)_6NHCO\right]_n$$

This reaction produces a highly crystalline polymer, but a vast number of urethanes can be formed by varying the reactants and a series of polymers covering a wide spectrum of properties can be prepared.

2.5 Stoichiometric control of M_n

Often it is preferable to avoid production of high molar mass polymer and control can be effected by rapidly cooling the reaction at the appropriate stage or by adding calculated quantities of monofunctional materials as in the preparation of nylon-6,6 from the salt.

More usefully, a precisely controlled stoichiometric imbalance of the reactants in the mixture can provide the desired result. For example, an excess of diamine over an acid chloride would eventually produce a polyamide with two amine end groups incapable of further growth when the acid chloride was totally con-sumed. This can be expressed in an extension of the Carothers equation as,

$$x_n = (1 + r)/(1 + r - 2rp),   \qquad (2.3)$$

where r is the ratio of the number of molecules of the reactants. Thus for a quantitative reaction ($p = 0.999$) between N molecules of phenolphthalein and $1.05N$ molecules of terephthaloyl chloride to form poly(terephthaloyl phenol-phthalein)

The value of $r = N_{AA}/N_{BB} = 1/1.05 = 0.952$

and $x_n = (1 + 0.952)/(1 + 0.952 - 2 \times 0.999 \times 0.952) \approx 39$,

rather than 1000 for $r = 1$. This reaction is an interfacial polycondensation whose progress may be followed by noting the colour change from the red phenolphthalein solution to the colourless polyester. The interface can remain stationary in the experiment but the uniformity of the polymer is improved by increasing the reaction surface using high-speed stirring.

The corresponding equation for a monofunctional additive is similar to equation (2.3) only now r is defined as the ratio $N_{AA}/(N_{BB} + 2N_B)$ where N_B is the number of monofunctional molecules added.

2.6 Kinetics
The assumption that functional group reactivity is independent of chain length can be verified kinetically by following a polyesterification. The simple esterification is an acid catalysed process where protonation of the acid is followed by interaction with the alcohol to produce an ester and water. If significant polymer formation is to be achieved, the water must be removed

continuously from the reaction to displace the equilibrium and the water eliminated can be used to estimate the extent of the reaction. Alternatively the rate of disappearance of carboxylic groups can be measured by titrating aliquots of the mixture.

A typical apparatus is shown in figure 2.2 consisting of a reaction kettle, a Dean and Stark trap to collect the water, a stirrer, N_2 inlet, and thermometer. The reaction can be illustrated using ethylene glycol and adipic acid. A mixture of

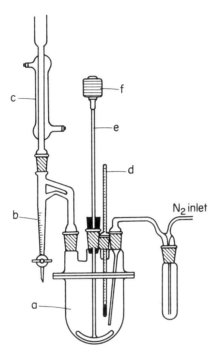

FIGURE 2.2. Apparatus suitable for polycondensation reactions: a, the reaction kettle; b, Dean and Stark trap; c, condenser; d, thermometer; e, stirrer; f, stirring motor.

decalin (35 cm^3) and adipic acid (1 mol) is placed in the kettle and the trap is filled with decalin. The mixture is heated to 420 K and glycol (1 mol), pre-heated to this temperature, is added, followed by an acid catalyst such as p-toluene sulphonic acid (1 mmol.). Nitrogen is bubbled rapidly through the mixture which is quickly raised to reflux temperature. The water level in the trap is noted at regular intervals and small aliquots of the reaction mixture can be withdrawn, weighed, diluted with acetone, and titrated with methanolic KOH. If an activation energy is required, the temperature can be raised several times and the reaction followed for a time at each new temperature.

Self-catalysed reaction. If no acid catalyst is added, the reaction will still proceed because the acid can act as its own catalyst. The rate of condensation at any time t can then be derived from the rate of disappearance of —COOH groups and

$$- d[COOH]/dt = k[COOH]^2[OH].$$ (2.4)

The second order [COOH] term arises from its use as a catalyst and k is the rate constant. For a system with equivalent quantities of acid and glycol the functional group concentration can be written simply as c and

$$- dc/dt = kc^3$$ (2.5)

This expression can be integrated under the conditions that $c = c_0$ at time $t = 0$ and

$$2kt = 1/c^2 - 1/c_0^2.$$ (2.6)

The water formed is removed and can be neglected. From the Carothers equation it follows that $c = c_0(1 - p)$, leading to the final form

$$2c_0^2 kt = 1/(1 - p)^2 - 1.$$ (2.7)

Acid-catalysed reaction. The uncatalysed reaction is rather slow and a high x_n is not readily attained. In the presence of a catalyst there is an acceleration of the rate and the kinetic expression is altered to

$$- d[COOH]/dt = k'[COOH][OH],$$ (2.8)

which is kinetically first order in each functional group. The new rate constant k' is then a composite of the rate constant k and the catalyst concentration which also remains constant. Hence

$$- dc/dt = k'c^2,$$ (2.9)

and integration gives finally

$$c_0 k't = 1/(1 - p) - 1.$$ (2.10)

Both equations have been verified experimentally by Flory as shown in figure 2.3.

2.7 Molar mass distribution in linear systems
The creation of long chain polymers by the covalent linking of small molecules is a random process, leading to chains of widely varying lengths. Because of the random nature of the process, the distribution of chain lengths in a sample can be arrived at by simple statistical arguments.

The problem is to calculate the probability of finding a chain composed of x basic structural units in the reaction mixture at time t, for either of the two reactions

$$x \text{ A–A} + x \text{ B–B} \rightarrow \text{A–A}[\text{B–BA–A}]_{x-1}\text{B–B},$$

$$\text{or } x \text{ A–B} \rightarrow \text{A}+\text{BA}+_{x-1}\text{B},$$

i.e. to calculate the probability that a functional group A or B has reacted. For the sake of clarity we can consider one of the functional groups to be carboxyl and determine the probability that $(x - 1)$ carboxyl groups have reacted to form a chain. This is p^{x-1}, where p is the extent of the reaction, defined in equation (2.1).

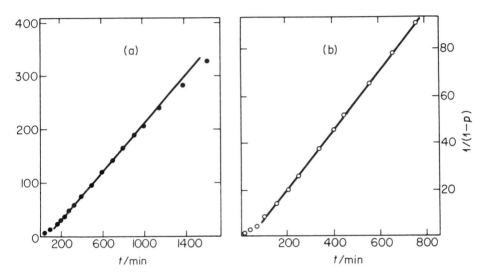

FIGURE 2.3. (a) Self-catalysed polyesterification of adipic acid with ethylene glycol at 439 K; (b) polyesterification of adipic acid with ethylene glycol at 382 K, catalysed by 0.4 mole per cent of *p*-toluene sulphonic acid. (From data by Flory.)

It follows that if a carboxyl group remains unreacted the probability of finding this uncondensed group is $(1 - p)$ and so the probability P_x of finding one chain x units long (*i.e.* an x-mer) is

$$P_x = (1 - p)p^{x-1}. \qquad (2.11)$$

As the fraction of x-mers in any system equals the probability of finding one, the total number N_x present is given by

$$N_x = N(1 - p)p^{x-1}, \qquad (2.12)$$

where N is the total number of polymer molecules present in the reaction. Substitution of the Carothers equation (2.2) gives

$$N_x = N_0(1 - p)^2 p^{x-1}, \qquad (2.13)$$

where N_0 is the total number of monomer units present initially. The variation of N_x for various values of p and x is shown in figure 2.4. A slightly different set of curves is obtained if the composition is expressed in terms of mass fraction w, in this case $w_x = xN_x/N_0$ to give

$$w_x = x(1-p)^2 p^{x-1}, \qquad (2.14)$$

Both reveal that very high conversions are necessary if chains of significant size are to be obtained and that while monomer is normally the most numerous species, the proportion of low molar mass material decreases as p exceeds 0.95.

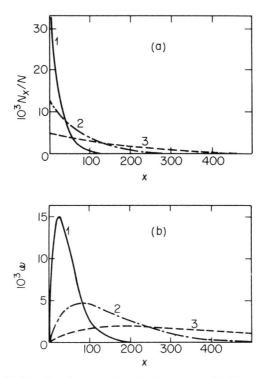

FIGURE 2.4. (a) Number fraction distribution curves for linear step growth polymerizations. Curve 1, $p = 0.9600$; Curve 2, $p = 0.9875$; Curve 3, $p = 0.9950$. (b) Corresponding weight fraction distributions for the same system.

2.8 Average molar masses

Number and weight average molar masses can be calculated from the equations if M_0 is taken as the molar mass of the repeat unit. Thus

$$M_n = N_x \Sigma(M_0 N_x)/N = M_0/(1-p), \qquad (2.15)$$

and

$$M_w = M_0(1+p)/(1-p). \qquad (2.16)$$

It can be seen that a heterogeneity index (M_w/M_n) for the most probable distribution when $p = 1$ is

$$(M_w/M_n) = \{M_0(1 + p)/(1 - p)\}\{(1 - p)/M_0\} = 2. \qquad (2.17)$$

2.9 Characteristics of step-growth polymerization

It might be appropriate at this stage to summarize the main features of the step-reactions.

(1) Any two molecular species in the mixture can react.

(2) The monomer is almost all incorporated in a chain molecule in the early stages of the reaction, i.e. about 1 per cent of monomer remains unreacted when $x_n = 10$. Hence polymer yield is independent of the reaction time in the later stages.

(3) Initiation, propagation, and termination reactions are essentially identical in rate and mechanism.

(4) The chain length increases steadily as the reaction proceeds.

(5) Long reaction times and high conversions are necessary for the production of a polymer with large x_n.

(6) Reaction rates are slow at ambient temperatures, but increase with a rise in temperature although this has little effect on the chain length of the final product.

(7) Activation energies are moderately high and reactions are not excessively exothermic.

2.10 Typical step-growth reactions

Reactions are normally carried out in bulk in the temperature range 420 to 520 K to encourage fast reactions and promote the removal of the low molar mass condensation product. Activation energies are about 80 kJ mol^{-1}.

Low temperature polycondensations. The advantages of high temperature reactions are partly counteracted by the increasing danger of side-reactions, and room temperature reactions using highly energetic reactants provide routes to a variety of polymers. The use of the Schotten–Baumann reaction for polyamides has already been outlined. This is an example of an unstirred interfacial reaction in which the diamine is soluble in both phases and diffuses across the interface into the organic layer where polymerization takes place. Continuous polymer production is achieved by withdrawing the film formed at the interface to allow continued diffusion of the reactants. Alternatively, the continuity of the polymerization reaction can be maintained by stirring the system vigorously; this ensures a constantly changing interface and increases the surface area available for the reaction. As both methods are diffusion controlled the need for stringent stoichiometric control is obviated.

When the diamine used is aromatic only low molar mass polymer is formed because of the lower reaction rates. To produce longer chains, conditions must be readjusted so that both phases are polar and miscible and vigorous high-speed stirring of the system is necessary. This is used in the reaction between iso-phthaloyl chloride and *m*-phenylene diamine,

These aromatic polyamides are particularly versatile materials and of considerable interest.

The ultimate extension to a homogeneous system with inert polar solvents is used in the synthesis of polyimides. Poly(methylene 4,4'-diphenylene pyro-mellitamide acid) is prepared by mixing equal amounts of pyromellitic dianhydride and bis(4-aminophenyl methane) in N,N'-dimethylformamide

The reaction is maintained at 288 K, with stirring, for an hour and the polymer is isolated by precipitation in vigorously stirred water.

Polysulphonamides, polyanhydrides, and polyurethanes can also be formed in homogeneous low temperature reactions.

Polyurethanes. An important and versatile family of polymers, whose diverse uses include foams, fibres, elastomers, adhesives, and coatings, is formed by the interaction of diisocyanates with diols to give chains with the characteristic $-(NHCOO)-$ link. Unfortunately the polymerizations are prone to side-reactions and an excess of the diisocyanate is often necessary initially to react with impurities in the system. Property variations in the product can be obtained through a suitable choice of starting materials. Thus high melting temperature compounds can be prepared if phenyl groups are incorporated in the chain by using biphenylene diisocyanate, whereas flexible elastomeric chains are formed if long $(CH_2)_x$ sequences are present. The side-reaction of the isocyanate group with water can be utilized to make foams. Liberation of CO_2

in a viscous polymerizing medium blows it up into an expanded cellular form; the rigidity or flexibility of the foam can be controlled by the chain flexibility of the polymer, and foam density by the amount of CO_2 liberated.

2.11 Ring formation
The assumption so far has been that all bifunctional monomers in step-growth reactions form linear polymers. This is not always true and competitive side reactions such as cyclization may occur, as with certain hydroxyacids which may form lactones, or lactams if an amino acid is used.

To gauge the importance of such reactions consideration must be given to the thermodynamic and kinetic aspects of ring formation. A study of ring strain in cycloalkanes has shown that 3 and 4 membered rings are severely strained but that this decreases dramatically for 5, 6, or 7 membered rings, then increases again up to 11 before decreasing for very large rings. In addition to the thermodynamic stability, the kinetic feasibility of two suitable functional groups being in juxtaposition to react must also be considered. This probability decreases with increasing ring size and again 5, 6, or 7 membered rings are favoured and will form in preference to a linear chain when possible.

2.12 Ring opening polymerizations

In contrast to ring formation, an important group of polymer producing reactions involve ring opening. The kinetics are similar to step-growth and the reactions are sometimes called step-addition, but in most cases a catalyst is required. An important example is the polymerization of ϵ-caprolactam to give nylon-6. The reaction is catalysed by water, at 520 K, but will also proceed in the presence of a strong base such as an alkali metal or metal amide. With HCl

$$\begin{array}{c}
CH_2-CH_2 \\
/ \qquad C=O \\
CH_2 \qquad | \\
\backslash \qquad N-H \\
CH_2-CH_2
\end{array} \qquad \rightarrow \qquad +NH(CH_2)_5CO+_n$$

as catalyst, a series of cyclic lactams showed the following reaction rates of linear polymerization: $8 > 7 > 11 \gg 5$ or 6 membered rings. Similarly lactones will polymerize when treated with aqueous alkaline solutions but the 5 and 6 membered rings are resistant to cleavage and polymer formation.

2.13 High temperature polymers

While many of the more common polymers are remarkably resistant to chemical attack and are stable when subjected to mechanical deformation, few can withstand the destructive effects of intense heat. Recent work has centred on the synthesis of chains incorporating: (i) thermally unreactive aromatic rings; (ii) resonance stabilized systems; (iii) crosslinked "ladder" structures; and (iv) protective side groups in an attempt to improve the thermal stability.

These approaches have met with considerable success. Aromatic polycarbonates, polyethers, and polysulphones are useful in situations where drastic temperature fluctuations are encountered but they are still only stable up to about 570 K and beyond this most fail.

For greater temperature stability different ring systems have been tried. Fibres from aromatic polyamides are capable of maintaining about 50 per cent of the ambient tensile strength when heated to 550 K. They also have the added advantage that they are extremely difficult to ignite and can be used to make excellent fire resistant clothing.

The polyimides

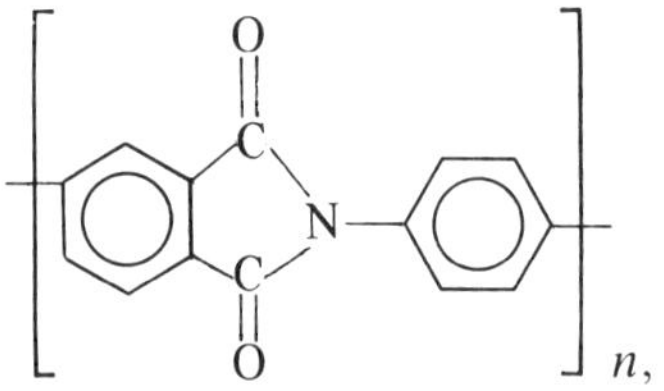

prepared in low temperature polycondensation reactions, constitute a further group some of whose members can withstand temperatures of up to 800 K. These are used in high temperature environments where lubricants cannot be used and are produced as sleeve bearings, valve seatings, and compressor vanes in jet engines. They are also excellent insulators but suffer from the disadvantage that they are normally intractable in common organic solvents and present manufacturing difficulties.

This problem is not met with in the polybenzimidazoles; these polymers are soluble yet retain their characteristic high temperature resistance and good mechanical response. They can be prepared in a two phase melt condensation reaction involving the diphenyl ester of a dibasic acid and a tetra-functional amine, *e.g.* polybenzimidazole,

This is an example of a semi-ladder polymer and the polybenzoxazoles

and polyquinoxalines

also fall into this category.

These structures derive their thermal stability from the high resonance energy of the aromatic and heterocyclic groups incorporated in the chain. In addition to these, polymeric azomethines with typical structures

show good thermal stability with negligible weight loss in air up to about 850 K.
They can be prepared by reacting an aryl dicarbonyl monomer

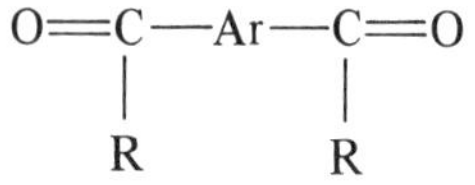

$$O{=}C{-}Ar{-}C{=}O$$
$$\quad|\qquad\quad|$$
$$\quad R\qquad\quad R$$

with a diamino aryl monomer. Alternatively a Schiff base exchange reaction
can be used where either of the two monomers will react with a Schiff base
such as $C_6H_5CR{=}NR'N{=}RCC_6H_5$.

Greater stability can be achieved if the vulnerable single bonds, which are
susceptible to degradation causing chain scission, can be eliminated. This can
be done by preparing a ladder structure, a typical example being poly(-imidazo-
pyrolone). It is easily seen that single bond scission at points A along the chain

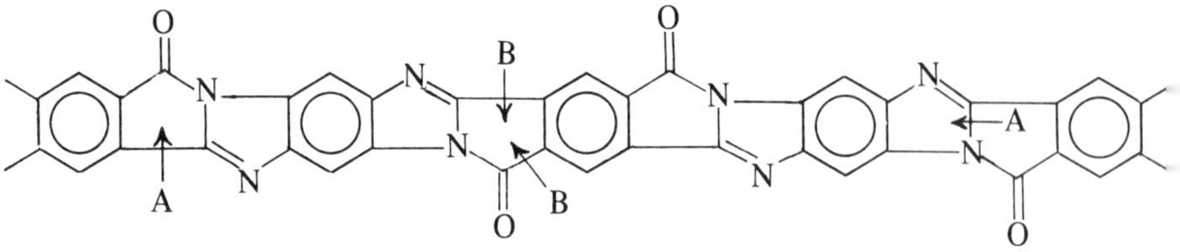

do not lead to complete chain scission. This can only be brought about by two
single bonds being broken as at points B, *i.e.* two bonds on opposite sides of
the chain and between the same two "rungs" of the ladder. As this process has
little probability many of these polymers can resist temperatures over 850 K
and so can compete with some metals as materials in this temperature range.

A number of these ladder polymers exist and an interesting one is the poly-
siloxane known as phenyl-T which combines the thermal stability of the ladder
structure with that arising from the use of inorganic elements instead of carbon.

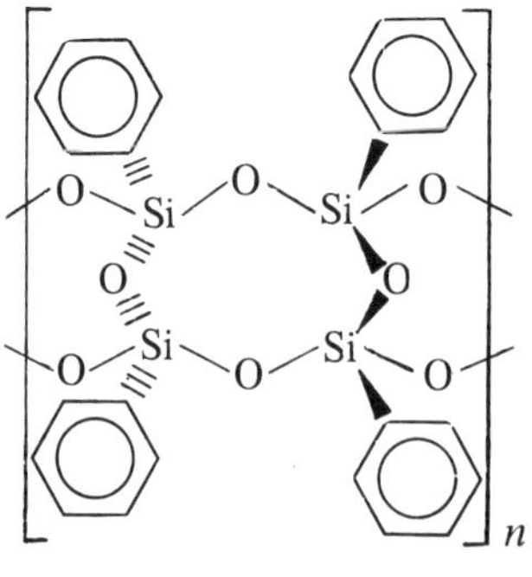

An alternative method of producing rigid chains is to prepare a spiro polymer.
In this structure one atom is common to two rings and an example of this is

the spiroketal obtained by condensing pentaerythritol with 1,4-cyclohexane-dione.

2.14 Non-linear step-growth reactions

In systems containing bifunctional monomers a high degree of polymerization is attained only when the reaction is forced almost to completion. The introduction of a trifunctional monomer into the reaction produces a rather startling change which is best illustrated using a modified form of the Carothers equation. A more general functionality factor f_{av} is introduced, defined as the average number of functional groups present per monomer unit. For a system containing N_0 molecules initially and equivalent numbers of two function groups A and B, the total number of functional groups is $N_0 f_{av}$. The number of groups that have reacted in time t to produce N molecules is then $2(N_0 - N)$ and

$$p = 2(N_0 - N)/N_0 f_{av}. \qquad (2.18)$$

The expression for x_n then becomes

$$x_n = 2/(2 - p f_{av}), \qquad (2.19)$$

but this is only valid when equal numbers of both functional groups are present in the system.

For a completely bifunctional system such as an equimolar mixture of phthalic acid and ethylene glycol, $f_{av} = 2$, and $x_n = 20$ for p = 0.95. If, however, a trifunctional alcohol, glycerol, is added so that the mixture is composed of 2 mol diacid, 1.4 mol diol, and 0.4 mol of glycerol f_{av} increases to

$$f_{av} = (2 \times 2 + 1.4 \times 2 + 0.4 \times 3)/3.8 = 2.1.$$

The value of x_n is now 200 after 95 per cent conversion but only a small increase to 95.23 per cent is required for x_n to approach infinity – a most dramatic increase. This is a direct result of incorporating a trifunctional unit in a linear chain where the unreacted hydroxyl provides an additional site for chain propagation. This leads to the formation of a highly branched structure and the greater the number of multi-functional units the faster the growth into an insoluble three-dimensional network. When this happens, the system is said to have reached its *gel point*.

If the stoichiometry of the system is unbalanced the definition of the average functionality must be modified and becomes

$$f' = 2rf_A f_B f_C / \{f_A f_C + r\rho f_A f_B + r(1 - \rho)f_B f_C\}, \qquad (2.20)$$

when monomers A and C have the same functional group but different functionalities f_A and f_C, and f_B is the functionality of monomer B. Also

$$r = (n_A f_A + n_C f_C)/n_B f_B \leqslant 1, \qquad (2.21)$$

$$\rho = n_C f_C/(n_A f_A + n_C f_C), \qquad (2.22)$$

and n_A, n_B and n_C are the amounts of each component. A commonly encountered system has monomers with $f_A = f_B = 2$ and $f_C > 2$. In this more general system the onset of gelation can be predicted by establishing the critical conversion limit beyond which a gel is sure to form. This is derived on the basis of the observation that x_n approaches infinity at or beyond the gel point. In the stoichiometric case the equation

$$p = (2/f_{av}) - (2/x_n f_{av}),$$

becomes at the gel point, when x_n tends to infinity,

$$p_G = 2/f_{av}, \qquad (2.23)$$

whereas the critical extent of reaction p_G for the general case is

$$p_G = (1 - \rho)/2 + 1/2r + \rho/f_C. \qquad (2.24)$$

2.15 Statistical derivation

An expression for p_G can be derived from statistical arguments. First a branching coefficient ζ is introduced, defined as the probability that a multifunctional monomer $(f > 2)$ is connected either by a linear chain segment or directly to a second multifunctional monomer (or branch point) rather than to a chain segment terminating in a single functional group.

The critical value ζ_G necessary for incipient gelation can be calculated from the probability that at least one of the $(f - 1)$ chain segments extending from a branch unit will be connected to another branch unit, and as this is simply $(f - 1)^{-1}$ we have

$$\zeta_G = 1/(f - 1), \qquad (2.25)$$

where f is now the functionality of the branching unit and not an average as defined previously. If more than one monomer in the system is multifunctional, then an average of all monomers with $f \geqslant 3$ is used.

It follows that if a chain ends in a branch point, $\zeta(f - 1)$ chains on the average will emanate from that point, and as ζ of these will also end in a branching unit a further $[\zeta(f - 1)]^2$ are generated and so on. This occurs when $\zeta(f - 1)$ is greater than 1 and the growth of the network is limited only by the boundaries of the system. For $\zeta(f - 1) < 1$ no gelation is observed.

The arguments can be extended and for a system A—A + B—B + A⧸A, the
critical extent of reaction of the A group at the gel point is

$$p_G = [r + r\rho(f - 2)]^{-1/2}. \qquad (2.26)$$

2.16 Comparison with experiment

The gel point in a branching system is usually detected by a rapid increase in
viscosity η, indicated by the inability of bubbles to rise in the medium. It is also
characterized by a rapid increase in x. The values of these quantities for the
reaction of diethylene glycol + succinic acid + 1,2,3,-propanetricarboxylic acid
are shown in figure 2.5. The increase in x_n is not as dramatic as x_w which can

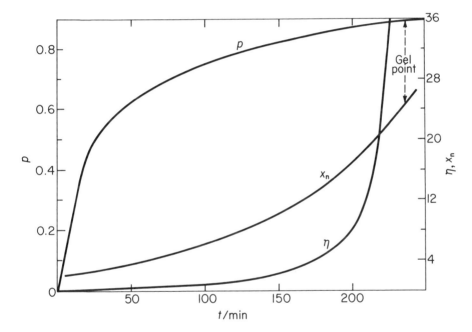

FIGURE 2.5. Variation of the viscosity η and the degree of polymerization x_n
with the extent of network polymer formation in the system diethylene
glycol + succinic acid + 1,2,3-propanetricarboxylic acid. (From data reported
by Flory.)

be identified with the η curve. The divergence of x_w and x_n is illustrated in
table 2.2 for the above system when the reaction mixture has been adjusted
hypothetically to provide a ratio of carboxyl to amine groups of 1. This
situation can be achieved by assuming that the mixture contains 98.5 mol of
diacid A, 100 mol of diol B, and 1 mol of triacid C. From equation (2.21) it
follows that $r = 98 \cdot 5 \times 2 + (1 \times 3)/(2 \times 100) = 1$. These reaction conditions
lead to the appearance of a gel point at $p_G = 0.9925$ and the ratio (x_w/x_n)

increases sharply as the reaction approaches the critical point. The distribution is readily compared with that for a totally bifunctional system for which $(x_w/x_n) = 1 + p$, and the broadening of the molar mass distribution, characteristic of these branching polymerizations is well illustrated.

TABLE 2.2. Branching system with $p_G = 0.9925$

p	x_w	x_w/x_n
0.100	1.2	1.1
0.500	3.0	1.5
0.700	5.8	1.7
0.900	20.4	2.0
0.950	45.6	2.2
0.980	153.8	2.7
0.990	747.2	5.6
0.992	3306.8	18.2

A comparison of experimental and theoretical values of p_G has been reported by Flory for a mixture of a tricarboxylic acid with a diol and a diacid; the results are given in table 2.3.

TABLE 2.3. 1,2,3-propanetricarboxylic acid, diethylene glycol, and adipic or succinic acid system

r	ρ	p_G		
		expt.	eqn.(2.24)	eqn.(2.26)
0.800	0.375	0.991	1.063	0.955
1.000	0.293	0.911	0.951	0.879
1.002	0.404	0.894	0.933	0.843

The statistical equation under estimates p_G whereas equation (2.24) over estimates the experimental value. The Carothers equation leads to a high p_G because molecules larger than the observed x_n exist in the mixture and these undergo gelation before the predicted value is attained. This difficulty is overcome in the statistical treatment but now differences are attributable to intra-molecular cyclization in the system and the loops which are formed are non-productive in a branching sense. This means that the reaction must proceed further to overcome waste.

2.17 Thermosetting polymers

The production of highly branched network polymers is commercially important, but as crosslinking results in a tough and highly intractable material, the fabrication process is usually carried out in two stages.

The first stage is the production of an incompletely reacted prepolymer; this is either solid or liquid and of moderately low molar mass. The second stage involves conversion of this into the final crosslinked product *in situ* (*i.e.* a

mould or form of some description). The prepolymers are either random or
structoset and are discussed separately.

Phenol-formaldehyde. Random prepolymers are prepared by reacting phenol
($f = 3$ for the *ortho* and *para* positions in the ring) with bifunctional formaldehyde. The base catalysed reaction produces a mixture of methylol phenols.

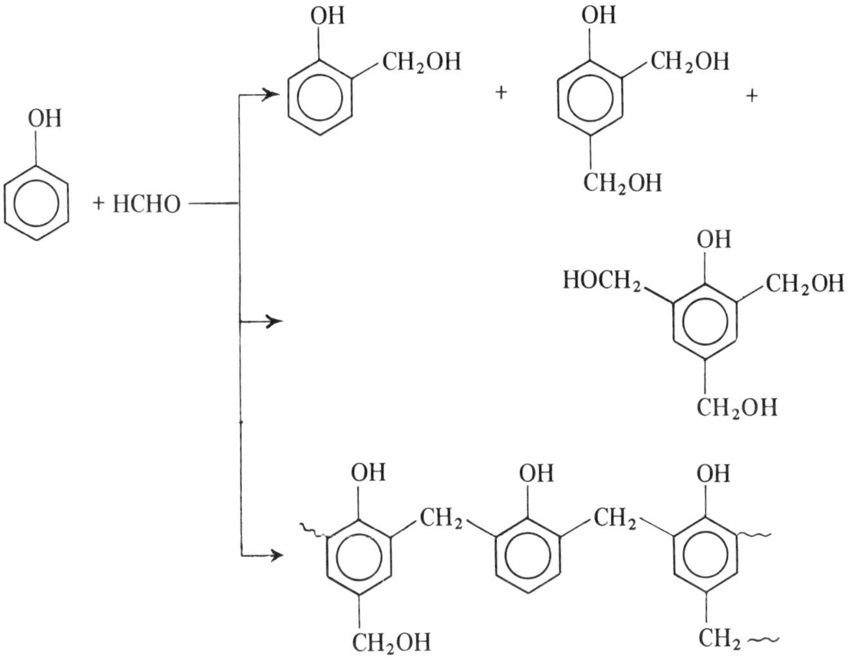

The composition of the mixture can be varied by altering the phenol to
formaldehyde ratio. At this stage the methylol intermediates are dried and
ground and in some cases a filler such as mica, glass fibre, or sawdust may be
added. A crosslinking agent such as hexamine is mixed with the prepolymer
together with CaO as a catalyst. When further heating takes place during
moulding, the hexamine decomposes to form HCHO and ammonia which
acts as a catalyst in the final crosslinking process by the HCHO.

Amino resins. A related family of polymers is made from random prepolymers
prepared by reacting either urea or melamine (shown below) with HCHO. The
products are known as aminoplasts.

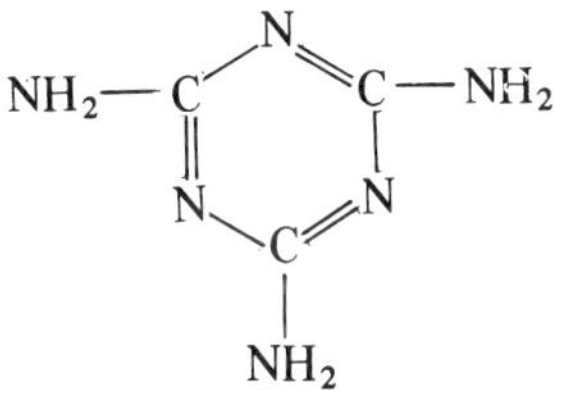

Epoxides. Structoset prepolymers are designed to have controlled and defined structures with functional groups located either at the chain ends — *structo-terminal* or located along the chain — *structopendant.* A prepolymer formed from bisphenol A and epichlorohydrin can be treated as either, depending on

$$
CH_2\!-\!CHCH_2Cl + HO\!-\!\!\langle\bigcirc\rangle\!-\!\underset{\underset{CH_3}{|}}{\overset{\overset{CH_3}{|}}{C}}\!-\!\langle\bigcirc\rangle\!-\!OH \rightarrow
$$

$$
CH_2\!-\!CHCH_2\!\left[\!-\!O\!-\!\langle\bigcirc\rangle\!-\!\underset{\underset{CH_3}{|}}{\overset{\overset{CH_3}{|}}{C}}\!-\!\langle\bigcirc\rangle\!-\!OCH_2\underset{\underset{OH}{|}}{C}HCH_2\!-\!\right]_n
$$

whether the functional site in the crosslinking is the pendant hydroxyl or the terminal epoxy. If crosslinking is achieved using a difunctional amine, the reactive site is the terminal epoxy group, whereas use of a polyanhydride leads to crosslinking through the −OH. Again fillers can be used to improve the properties of the final product. These are widely employed in tool making, adhesives, insulators, and tough surface coatings.

General Reading

P. J. Flory, *Principles of Polymer Chemistry,* Chapter 3. Cornell University Press, Ithaca, N.Y. (1953).

A. H. Frazer, *High Temperature Resistant Polymers.* Interscience (1968).

R. W. Lenz, *Organic Chemistry of Synthetic High Polymers,* Chapter 3. Interscience publishers Inc. (1967).

H. F. Mark and G. S. Whitby, *The Collected Papers of Wallace Hume Carothers.* Interscience Publishers Inc. (1940).

P. W. Morgan, *Condensation Polymers by Interface and Solution Methods.* Interscience Publishers Inc. (1965).

J. W. Mulvaney, "Heat-resistant polymers" in *Encyclopaedia of Polymer Science and Technology,* Vol. 7. Interscience Publishers Inc. (1967).

G. Odian, *Principles of Polymerization,* Chapter 2. McGraw-Hill (1970).

J. H. Saunders and K. C. Frisch, *Polyurethanes, Chemistry and Technology,* Vols. I and II. Interscience Publishers Inc. (1964).

R. B. Seymour, *Introduction to Polymer Chemistry,* Chapter 5. McGraw-Hill (1971).

W. M. Smith, *Manufacture of Plastics,* Chapters 9 and 13. Reinhold Publishing Corp. (1964).

D. H. Solomon, *The Chemistry of Organic Film Formers.* John Wiley and Sons, Inc. (1967).

References

1. P. J. Flory, (a) *J. Amer. Chem. Soc., 61,* 3334 (1939);
          (b) **62,** 2261 (1940);
          (c) **63,** 3083 (1941).

CHAPTER 3

Free Radical Addition Polymerization

3.1 Addition polymerization

In step-growth polymerization reactions it is often necessary to use multifunctional monomers if polymers with high molar masses are to be formed; this is not the case when addition reactions are employed. Long chains are readily obtained from monomers such as vinylidene compounds with the general structure $CH_2 = CR_1R_2$. These are bi-functional units, where the special reactivity of π-bonds in the carbon to carbon double bond makes them susceptible to rearrangement if activated by free radical or ionic initiators. The active centre created by this reaction then propagates a kinetic chain which leads to the formation of a single macromolecule whose growth is stopped when the active centre is neutralized by a termination reaction. The complete polymerization proceeds in three distinct stages: (i) *Initiation,* when the active centre which acts as a chain carrier is created; (ii) *Propagation*, involving growth of the macro-molecular chain by a kinetic chain mechanism and characterized by a long sequence of identical events, namely the repeated addition of a monomer to the growing chain; (iii) *Termination,* whereby the kinetic chain is brought to a halt by the neutralization or transfer of the active centre. Typically the polymer formed has the same chemical composition as the monomer, *i.e.* each unit in the chain is a complete monomer and not a residue as in most step-growth reactions.

3.2 Choice of initiators

A variety of chain initiators are available to the polymer chemist and fall into three general categories: free radical; cationic; and anionic. The choice of the most appropriate one depends largely on the groups R_1 and R_2 in the monomer and their effect on the double bond. This arises from the ability of the alkene

π-bond to react in a different way with each initiator species to produce either

$$\overset{+}{\underset{|}{\overset{|}{C}}}-\overset{|}{\underset{|}{\overset{-}{C}}} \quad \rightleftharpoons \quad \overset{|}{\underset{|}{C}}=\overset{|}{\underset{|}{C}} \quad \rightleftharpoons \quad \overset{\cdot}{\underset{|}{\overset{|}{C}}}-\overset{|}{\underset{|}{\overset{\cdot}{C}}}$$

$$\qquad\quad \text{I} \qquad\qquad\qquad\qquad\qquad\qquad\qquad \text{II}$$

heterolytic (I) or homolytic (II) fission. In most olefinic monomers of interest the group R_1 is either H or CH_3 and for simplicity we can consider it to be H. The group R_2 is then classifiable as an electron withdrawing group $CH_2 = \overset{\delta+}{C}\overset{\delta-}{H} \rightarrow \overset{}{R_2}$ or an electron donating group $\overset{\delta-}{CH_2} = CH \leftarrow \overset{\delta+}{R_2}$. Both alter the negativity of the π-bond electron cloud and thereby determine whether or not a radical, an anion, or a cation will be stabilized preferentially.

In general, electron withdrawing substituents, $-CN$, $-COOR$, $-CONH_2$, reduce the electron density at the double bond and favour propagation by an anionic species. Groups which tend to increase the double-bond nucleophilicity by donating electrons, such as alkenyl, alkoxyl, and phenyl, encourage attack by cationic initiators and in addition the active centres formed are resonance stabilized. Alkyl groups do not stimulate cationic initiation unless in the form of $1,1'$-dialkyl monomers or alkyl dienes and then heterogeneous catalysts are necessary. As resonance stabilization of the active centre is an important factor, monomers like styrene and 1,3-butadiene can undergo polymerization by both ionic methods because the anionic species can also be stabilized.

Because of its electrical neutrality, the free radical is a less selective and more generally useful initiator because most substituents can provide some resonance stabilization for this propagating species. Some examples are shown in table 3.1.

TABLE 3.1. Effect of substituent on choice of initiator

Monomer	Initiator		
	Free radical	Anionic	Cationic
Ethylene, $CH_2 = CH_2$	+	−	+
$1,1'$-Dialkylolefin, $CH_2 = CR_1 R_2$	−	−	+
Vinyl ethers, $CH_2 = CHOR$	−	−	+
Vinyl halides, $CH_2 = CH(Hal)$	+	−	−
Vinyl esters, $CH_2 = CHOCOR$	+	−	−
Methacrylic esters, $CH_2 = C(CH_3)COOR$	+	+	−
Acrylonitrile, $CH_2 = CHCN$	+	+	−
Styrene, $CH_2 = CHPh$	+	+	+
1,3-Butadiene, $CH_2 = CR.\ CH = CH_2$	+	+	+

3.3 Free radical polymerization

A free radical is an atomic or molecular species whose normal bonding system has been modified such that an unpaired electron remains associated with the new structure. The radical is capable of reacting with an olefinic monomer to generate a chain carrier which can retain its activity long enough to propagate a macromolecular chain under the appropriate conditions.

$$R^{\cdot} + CH_2 = CHR_1 \rightarrow RCH_2CHR_1^{\cdot}$$

3.4 Initiators

An effective initiator is a molecule which, when subjected to heat, electromagnetic radiation, or chemical reaction, will readily undergo homolytic fission into radicals of greater reactivity than the monomer radical. These radicals must also be stable long enough to react with a monomer and create an active centre. Particularly useful for kinetic studies are compounds containing an azonitrile group as the decomposition kinetics are normally first order and the rates are unaffected by the solvent environment.

Typical radical producing reactions are:

(1) *Thermal decomposition* can be usefully applied to organic peroxides or azo compounds, *e.g.* benzoyl peroxide when heated eventually forms two phenyl radicals with loss of CO_2. A simpler one-stage decomposition is obtained when

dicumyl peroxide is used.

$$C_6H_5\!-\!C(CH_3)_2\!-\!O\!-\!O\!-\!(CH_3)_2C\!-\!C_6H_5 \rightarrow 2C_6H_5\!-\!C(CH_3)_2O^{\cdot}$$

(2) *Photolysis* is applicable to metal iodides, metal alkyls, and azo compounds, *e.g.* α,α'-azobisisobutyronitrile is decomposed by radiation with a wavelength of 360 nm.

Note that in each reaction there are two radicals $R^{\cdot}$ produced from one initiator molecule I; in general:

$$I \rightarrow 2R^{\cdot} \tag{3.1}$$

(3) *Redox reactions, e.g.* the reaction between the ferrous ion and hydrogen peroxide in solution produces hydroxyl radicals,

$$H_2O_2 + Fe^{2+} \rightarrow Fe^{3+} + OH^- + OH^\bullet.$$

Alkyl hydroperoxides may be used in place of H_2O_2. A similar reaction is observed when cerium(IV) sulphate oxidizes an alcohol:

$$RCH_2OH + Ce^{4+} \rightarrow Ce^{3+} + H^+ + RC(OH)H^\bullet.$$

(4) *Persulphates* are useful in emulsion polymerizations where decomposition occurs in the aqueous phase and the radical diffuses into a hydrophobic, monomer containing, droplet.

$$S_2O_8^{2-} \rightarrow 2S\dot{O}_4^-$$

(5) *Ionizing radiation* such as α-, β-, γ-, or X-rays may be used to initiate a polymerization, by causing the ejection of an electron followed by dissociation and electron capture to produce a radical.

Ejection: $C \rightsquigarrow C^+ + e^-$

Dissociation: $C^+ \longrightarrow A^\bullet + Q^+$

e^--Capture: $Q^+ + e^- \longrightarrow Q^\bullet$

Useful initiators in polymerizations are compounds providing a source of free radicals in the temperature range 320 to 420 K and the half-lives of several of the more common are shown in figure 3.1 as a function of temperature. Reference to these data provides the chemist with the information needed to make a suitable selection for the chosen polymerization conditions.

INITIATOR EFFICIENCY

Although the decomposition of an initiator molecule can be quantitative, chain initiation may be less than 100 per cent efficient. In a kinetic analysis the effective radical concentration is represented by an *efficiency factor f* which is less than unity when only a proportion of the radicals generated are effective in the creation of a kinetic chain. Inefficient chain propagation may be due to several side reactions.

Primary recombination can occur if the diffusion of radical fragments is impeded in the solution and a cage effect leads to

$$2(CH_3)_2C^\bullet \rightarrow (CH_3)_2C \text{---} C \quad (CH_3)_2 .$$
$$ | \qquad\qquad | \quad |$$
$$ CN \qquad\quad CN \quad CN$$

The solvent usually plays an important part and the extent of the decomposition of benzoyl peroxide is limited to 35 per cent in tetrachloroethylene, 50 per cent in benzene, and 85 per cent in ethyl acetate, after refluxing for 4 h.

Further wastage occurs when induced decomposition is effected by the attack of an active centre

$$R^{\bullet} + R'\!-\!O\!-\!O\!-\!R' \rightarrow ROR' + R'O^{\bullet}$$

This effectively produces one radical instead of the three potential radical species.

For a 100 per cent efficient initiator $f = 1$, but most initiators have efficiencies in the range 0.3 to 0.8.

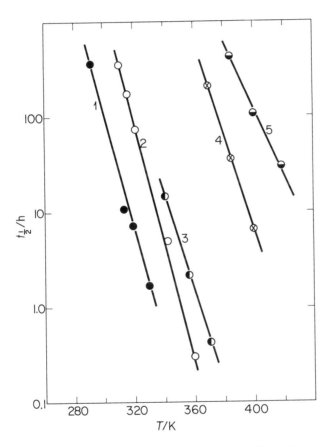

FIGURE 3.1. Half-lives $t_{\frac{1}{2}}$ of selected peroxide initiators. Curve 1, isopropyl percarbonate; curve 2, 2,2'-azo-bis-isobutyronitrile; curve 3, benzoyl peroxide; curve 4, di-tertiary butyl peroxide; curve 5, cumene hydroperoxide.

3.5 Chain growth
A chain carrier is formed from the reaction of the free radical and a monomer

unit; chain propagation then proceeds rapidly by addition to produce a linear polymer.

$$RM_1^{\bullet} + M_1 \rightarrow RM_2^{\bullet} \qquad (3.2)$$

$$RM_n^{\bullet} + M_1 \rightarrow RM_{n+1}^{\bullet} \qquad (3.3)$$

The average life time of the growing chain is short, but a chain of over 1000 units can be produced in 10^{-2} to 10^{-3} s. Bamford and Dewar have estimated that the thermal polymerization of styrene at 373 K leads to chains of $x = 1650$ in approximately 1.24 s, *i.e.* a monomer adds on once every 0.75 ms.

3.6 Termination

In theory the chain could continue to propagate until all the monomer in the system had been consumed but for the fact that free radicals are particularly reactive species and interact as quickly as possible to form inactive covalent bonds. This means that short chains are produced if the radical concentration is high, because the probability of radical interaction is correspondingly high, and the radical concentration should be kept small if long chains are required. Termination of chains can take place in several ways: (1) the interaction of two active chain ends; (2) the reaction of an active chain end with an initiator radical; (3) termination by transfer of the active centre to another molecule which may be solvent, initiator, or monomer; (4) interaction with impurities (*e.g.* oxygen) or inhibitors.

The most important termination reaction is the first, a bimolecular interaction between two chain ends. Two routes are possible.

(a) *Combination* where two chain ends couple together to form one long chain.

$$\sim CH_2{-}\overset{\displaystyle |}{\underset{\displaystyle Cl}{CH}}^{\bullet} + HC^{\bullet}\overset{\displaystyle |}{\underset{\displaystyle Cl}{{-}}}CH_2\sim \rightarrow \sim CH_2{-}\overset{\displaystyle |}{\underset{\displaystyle Cl}{CH}}{-}\overset{\displaystyle |}{\underset{\displaystyle Cl}{CH}}{-}CH_2\sim$$

(b) *Disproportionation* with hydrogen abstraction from one end to give an unsaturated group and two dead polymer chains.

$$\sim CH_2{-}\overset{\displaystyle CH_3}{\underset{\displaystyle COOCH_3}{\overset{\displaystyle |}{\underset{\displaystyle |}{C}}}}{}^{\bullet} + {}^{\bullet}\overset{\displaystyle CH_3}{\underset{\displaystyle COOCH_3}{\overset{\displaystyle |}{\underset{\displaystyle |}{C}}}}{-}CH_2\sim \rightarrow \sim CH{=}\overset{\displaystyle CH_3}{\underset{\displaystyle COOCH_3}{\overset{\displaystyle |}{\underset{\displaystyle |}{C}}}} + \overset{\displaystyle CH_3}{\underset{\displaystyle COOCH_3}{\overset{\displaystyle |}{\underset{\displaystyle |}{CH}}}}{-}CH_2\sim$$

One or both processes may be active in any system depending on the monomer and polymerizing conditions. Experimental evidence suggests that polystyrene terminates predominantly by combination whereas poly(methyl methacrylate) terminates exclusively by disproportionation when the reaction is above 333 K but by both mechanisms below this temperature. The mechanism can be determined by measuring the number of initiator fragments per chain using a radioactive initiator. One fragment per chain is counted when disproportionation is operative and two when combination occurs. Alternatively the number average molar mass of the product can be measured.

3.7 Steady-state kinetics

The three basic steps in the polymerization process can be expressed in general terms as follows:

Initiation is a two stage reaction. Initiator decomposition

$$I \xrightarrow{k_d} 2R^{\bullet},\qquad(3.4)$$

is followed by radical attack on a monomer unit to form a chain carrier,

$$R^{\bullet} + M \xrightarrow{k_i} RM^{\bullet}.\qquad(3.5)$$

Since the initial decomposition is slow compared with both the rate of addition of a primary radical to a monomer and the termination reaction, it is the rate determining step. The rate of initiation v_i is then the rate of production of chain radicals

$$v_i = d[RM^{\bullet}]/dt = 2k_d f[I],\qquad(3.6)$$

where the factor 2 is introduced because two potentially effective radicals are produced in the decomposition; f also measures the ability of these to propagate chains. This expression is valid for a thermo-initiation, but many reactions can be photo-initiated when the monomer absorbs radiation and acts as its own initiator. The rate v_{ip} is then dependent on the intensity of light absorbed

$$v_{ip} = 2\phi I_a.\qquad(3.7)$$

The quantum yield ϕ replaces f and defines the initiator efficiency; I_a is related to the incident light intensity I_0, the monomer concentration, and the extinction coefficient ϵ so that

$$v_{ip} = 2\phi\epsilon I_0 [M].\qquad(3.8)$$

When the monomer is a poor absorber of radiation small quantities of a photosensitizer may be added to absorb the energy and then transfer this to the monomer to create an active centre. In this case [M] is replaced by the concentration of photosensitizer.

Propagation is the addition of monomer to the growing radical

$$RM_n^{\bullet} + M_1 \xrightarrow{k_p} RM_{n+1}^{\bullet}\qquad(3.9)$$

The rate of bimolecular propagation is assumed to be the same for each step so that

$$v_p = k_p [M] [M^{\bullet}],\qquad(3.10)$$

where $[M^{\bullet}]$ represents the concentration of growing ends; and $[M^{\bullet}]$ is usually low at any particular time. The reaction is essentially the conversion of monomer to polymer and can be followed from the rate of disappearance of monomer.

Termination is also a bimolecular process depending only on $[M^\bullet]$; v_t for both mechanisms is

$$v_t = 2k_t[M^\bullet][M^\bullet]. \quad (3.11)$$

The rate constant k_t is actually $(k_{tc} + k_{td})$ where the two mechanisms, combination and disproportionation are possible but will be written as k_t for convenience. If the chain reaction does not lead to an explosion, a steady state is reached where the rate of radical formation is exactly counterbalanced by the rate of destruction, *i.e.* $v_i = v_t$, and for a thermal reaction

$$2k_t[M^\bullet]^2 = 2k_d f[I]. \quad (3.12)$$

From this an expression for $[M^\bullet]$ is obtained

$$[M\bullet] = \left\{ fk_d[I]/k_t \right\}^{1/2}, \quad (3.13)$$

and because the concentration of radicals is usually too low to be determined accurately it is replaced in the kinetic expression. The overall rate of polymerization is then

$$v_p = k_p \left\{ fk_d[I]/k_t \right\}^{1/2}[M], \quad (3.14)$$

which shows that the rate is proportional to the monomer concentration and to $[I]^{1/2}$ if f is high, but for a low efficiency initiator f becomes a function of $[M]$ and the rate is then proportional to $[M]^{3/2}$.

By analogy the rate of a photo-polymerization becomes

$$v_{pp} = k_{pp} \left\{ \phi \epsilon I_0/k_t \right\}^{1/2}[M]^{3/2}. \quad (3.15)$$

Two parameters of interest can be derived from this analysis, the kinetic chain length $\bar{v}$ and the average degree of polymerization x.

The kinetic chain length $\bar{v}$ is a measure of the average number of monomer units reacting with an active centre during its lifetime and is related to x_n through the mechanism of the termination. Thus combination means $x_n = 2\bar{v}$ but $x_n = \bar{v}$ if disproportionation is the only termination reaction. Under steady-state conditions

$$\bar{v} = v_p/v_i = v_p/v_t = k_p^2[M]^2/2k_t v_p. \quad (3.16)$$

This means that as $\bar{v}$ is inversely proportional to the rate of polymerization, an increase in temperature produces an increase in v_p and a corresponding decrease in chain length. The equation shows that $\bar{v}$ is inversely proportional to the radical concentration, hence the chain length is short for high radical concentrations and vice-versa. This means that a certain degree of control over the polymer chain length can be exercised by altering the initiator concentration.

This kinetic analysis is an oversimplification based as it is on the assumptions that termination occurs solely by a reaction between two growing chains, that no chain transfer takes place, that a head-to-tail arrangement of monomer units is uniform throughout the chain, and that there is negligible auto-acceleration.

3.8 Trommsdorff-Norrish effect

In many polymerizations a marked increase in rate is observed towards the end of the reaction instead of the expected gradual decrease caused by the depletion of the monomer and initiator. This *auto-acceleration* is a direct result of the increased viscosity of the medium and the effect is most dramatic when polymerizations are carried out in the bulk phase or in concentrated solutions. The phenomenon, sometimes known as the *Trommsdorff-Norrish* or *gel effect,* is caused by the loss of the steady state in the polymerization kinetics. Neither the initiation nor the propagation are diffusion controlled reactions in the early stages because they only require the migration of a monomer to the active centre. Termination, however, is a bimolecular process requiring fruitful collision between two radical ends attached to long and highly entangled polymer chains. When the viscosity of the medium is high, extensive entanglements hinder the movement of the active chain end into a position where it can react with another radical end. This can have a profound effect on the termination rate constant k_t and may reduce it by one or perhaps two orders of magnitude, while leaving k_d and k_p relatively unaffected. This means that the rate, proportional to $(k_p k_d^{1/2}/k_t^{1/2})$ will increase and sufficient energy may also be liberated to produce an explosion if its dissipation is poor.

The auto-acceleration can be avoided by performing the polymerization in more dilute solutions or by stopping the reaction before the diffusion effect grows to noticeable proportions. However, polymerizations in solvents can be influenced by the choice of medium and this feature is treated in the chain transfer reactions.

3.9 Chain transfer

Termination in a free radical polymerization normally occurs by collision between two active centres attached to polymer chains, but the chain length of the product in many systems is lower than one would expect if this was the mechanism solely responsible for limiting the kinetic chain length $\bar{\nu}$. Usually x_n will lie within the expected limits of $\bar{\nu}$ (disproportionation) and $2\bar{\nu}$ (combination), but not always, and Flory found that attenuation of chain growth takes place if there is premature termination of the propagating chain by a transfer of activity to another species through a collision. This is a competitive process involving the abstraction of an atom by a chain carrier from an inactive molecule XY with replaceable atoms, and is dependent on the strength of the X–Y bond.

$$\sim M_m^{\,\cdot} + XY \rightarrow \sim M_m X + Y^{\,\cdot}$$

It is important to note that the free radical is not destroyed in the reaction, merely transferred, and if the new species is sufficiently active another chain will emanate from the new centre. This is known as *chain transfer* and is a reaction resulting in the exchange of an active centre between molecules during a bimolecular collision. Several types of chain transfer have been identified.

Transfer to monomer. The two important reactions in this group both involve hydrogen abstraction. Two competitive alternatives exist in the first group

$$R^\bullet + CH_2{=}CHX \begin{cases} \to RH + CH_2{=}C^\bullet X & \text{(I)} \\ \to RCH_2CHX^\bullet & \text{(II)} \end{cases}$$

If the radical formed in reaction (II) is virtually unstabilized by resonance, then the reaction with the parent unreactive monomer may produce little chain propagation due to the tendency for stabilization to occur by removal of hydrogen from the monomer. This leads to rapid chain termination and is known as *degradative transfer.* Allylic monomers are particularly prone to this type of reaction,

$$\sim R^\bullet + CH_2{=}CHCH_2OCOCH_3 \to \sim RH + \overset{\bullet}{C}H_2{\cdots}CH{\cdots}\overset{\bullet}{C}H{-}OCOCH_3,$$

where abstraction of the α-hydrogen leads to a resonance stabilized allylic radical capable only of bimolecular combination with another allyl radical. This is effectively an auto-inhibition by the monomer. Propylene also reacts in this manner and both monomers are reluctant to polymerize by a free radical mechanism.

A second group of transfer reactions can occur by hydrogen abstraction from the pendant group.

The relevant kinetic expression is

$$v_{tr} = k_{tr}^M [M] [M^\bullet]. \tag{3.17}$$

Transfer to initiator. Organic peroxides, when used as initiators, are particularly susceptible to chain transfer. Azo initiators are not vulnerable in this respect and are more useful when a kinetic analysis is required. For peroxides

$$v_{tr} = k_{tr}^I [I] [M^\bullet]. \tag{3.18}$$

Transfer to polymer. The transfer reaction with a polymer chain leads to branching rather than initiation of a new chain so that the average molar mass is relatively unaffected. The long and short chain branching detected in polyethylene is believed to arise from this mode of transfer.

Transfer to modifier. Molar masses can be controlled by addition of a known and efficient chain transfer agent such as an alkyl mercaptan.

$$\sim CH_2CHX^\bullet + RSH \to \sim CH_2CH_2X + RS^\bullet$$

$$RS^\bullet + CH_2{=}CHX \to RSCH_2CHX^\bullet$$

Mercaptans are commonly used because the S—H bond is weaker and more susceptible to chain transfer than a C—H bond.

Transfer to solvent. A significant decrease in polymer chain length is often found when polymerizations are carried out in solution rather than in the undiluted state and this variation is a function of both the extent of dilution and the type of solvent used. The effectiveness of a solvent in a transfer reaction depends largely on the amount present, the strength of the bond involved in the abstraction step, and the stability of the solvent radical formed. With the exception of fluorine, halogen atoms are easily transferred and the reaction of styrene in CCl_4 is a good example of this chain transfer.

$$\sim CH_2-\overset{\displaystyle |}{\underset{\displaystyle C_6H_5}{CH}}^{\displaystyle \cdot} + CCl_4 \rightarrow \sim CH_2\overset{\displaystyle |}{\underset{\displaystyle C_6H_5}{CHCl}} + CCl_3^{\cdot} \qquad (I)$$

$$CCl_3^{\cdot} + CH_2{=}CHC_6H_5 \rightarrow Cl_3CCH_2-\overset{\displaystyle |}{\underset{\displaystyle C_6H_5}{CH}}^{\displaystyle \cdot} \qquad (II)$$

$$Cl_3C\sim \overset{\displaystyle |}{\underset{\displaystyle C_6H_5}{CH}}^{\displaystyle \cdot} + CCl_4 \rightarrow Cl_3C\sim \overset{\displaystyle |}{\underset{\displaystyle C_6H_5}{CHCl}} + CCl_3^{\cdot} \qquad (III)$$

When the solvent is present in significant quantity, step (1) is of minor importance and the resulting polymer contains four chlorine atoms which can be detected by analysis.

Hydrogen is normally the atom abstracted and as radical stability enhances the transfer reaction, we find that toluene, which forms a primary radical, is less efficient than ethyl benzene, which forms a secondary radical, while both are inferior to isopropyl benzene which forms a tertiary radical. All are much better than *t*-butyl benzene, whose radical is unstable, so that virtually no chain transfer takes place in this solvent. It is interesting to note that even benzene acts as a chain transfer agent on a modest scale.

The kinetic expression is

$$v_{tr} = k_{tr}^S [M] [S] . \qquad (3.19)$$

CONSEQUENCES OF CHAIN TRANSFER

The primary effect is a decrease in the polymer chain length, but other less obvious occurrences can be detected. If k_{tr} is much larger than k_p then a very small polymer is formed with x_n between 2 and 5. This is known as *telomeriza-tion*. The chain re-initiation process can also be slower than the propagation reaction and a decrease in v_p is observed. However, the influence on x_n is most important and it can be estimated by considering all the transfer processes in a form known as the Mayo equation.

$$1/x_n = (1/x_n)_0 + C_s[S]/[M] \qquad (3.20)$$

This is a simplified form in which the main assumption is that solvent transfer predominates and all other terms are included in $(1/x_n)_0$. The chain transfer constant C_s is then (k_{tr}^s/k_p).

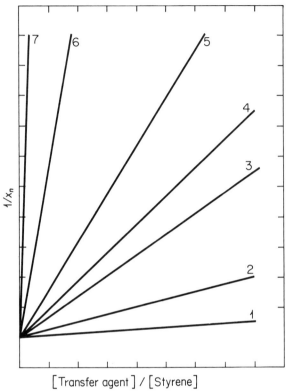

[Transfer agent] / [Styrene]

FIGURE 3.2. Effect of chain transfer to various solvents on the degree of polymerization of polystyrene at 333 K. 1, Benzene; 2, n-Heptane; 3, sec-Butyl benzene; 4, m-Cresol; 5, CCl_4; 6, CBr_4; 7, n-Butyl mercaptan.

A plot of $1/x_n$ against $\{[S]/[M]\}$ for a variety of agents is shown in figure 3.2. The slope is a measure of C_s and the intercept is $(1/x_n)_0$. If the activation energy of the process is required, $\log C_s$ can be plotted against $1/T$.

TABLE 3.2. Chain transfer constants of various agents to styrene at 333 K

Agent	$10^4 C_s$
Benzene	0.023
n-Heptane	0.42
sec-Butyl benzene	6.22
m-Cresol	11.0
CCl_4	90
CBr_4	22 000
n-Butylmercaptan	210 000

3.10 Inhibitors and retarders

Chain transfer agents can lower the average chain length and in extreme cases, when used in large proportions, may lead to the formation of telomers.

Some chain transfer agents yield radicals with low activity and if the re-initiation reaction is slow, the polymerization rate decreases because there is a build-up of radicals leading to increased termination by coupling. When this happens the substance responsible is said to be a *retarder, e.g.* nitrobenzene acts in this way with styrene.

In extreme cases an added reagent may suppress polymerization completely by reacting with the initiating radical species and converting them all efficiently into unreactive substances. This is known as inhibition, but the difference between an inhibitor and a retarder is merely in the degree of efficiency.

The phenomena are typified by the reaction of styrene with benzoquinone, nitrobenzene, and nitrosobenzene, studied by Schulz, and shown in figure 3.3.

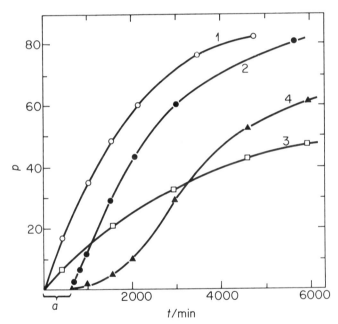

FIGURE 3.3 Polymerization of styrene at 373 K in the presence of: curve 1, no inhibitor; curve 2, 0.1 per cent benzoquinone, curve 3, 0.5 per cent nitrobenzene; curve 4, 0.2 per cent nitrosobenzene. The time t and percentage conversion p are plotted.

Curve 1 represents the polymerization of styrene in the absence of any agents. When benzoquinone is added the polymerization is completely inhibited until all the benzoquinone has been consumed, then the reaction proceeds normally (curve 2). The time interval a is the *induction period* and reflects the time taken for the benzoquinone to react with all the radicals formed until no more benzoquinone is left unreacted. In the presence of nitrobenzene, curve 3, the polymerization continues but at a much reduced rate. The action of nitroso-benzene is much more complex (curve 4). It acts first as an inhibitor but probably produces a substance during this period which then acts as a retarder and both effects are observed.

Monomers are usually transported and stored in the presence of inhibitors to prevent premature polymerization, and so must be redistilled and purified prior to use.

An excellent inhibitor is the resonance stabilized radical diphenyl picryl hydrazyl (DPPH), used extensively as a radical scavenger because the stoichiometry of the reaction is 1:1.

3.11 Experimental determination of individual rate constants

The parameters of interest are f, k_d, k_p, and k_t. Both k_d and f can be measured without resorting to polymerizations, but the steady-state kinetic scheme does not allow direct determination of the individual constants k_p and k_t, only the ratio (k_p^2/k_t). To separate these, non-steady-state conditions must be used, but first k_d and f can be measured.

Initiator decomposition and efficiency. An initiator is usually a compound with a labile bond whose dissociation energy lies in the range 105 to 170 kJ mol^{-1}. As the rate of decomposition of the initiator is normally the rate determining step this must be measured if v_i is required. The thermal dissociation of α,α'-azobisisobutyronitrile can be followed, in the absence of monomer, by following the rate of evolution of nitrogen during radical formation; a typical value for k_d is 1.2×10^{-5} s^{-1} at 333 K.

Thermo-initiated polymerizations suffer from the disadvantage that because of the large heat capacity of the system radical generation is difficult to control; also k_d measured in the absence of monomer may be invalid. Photo-initiation, where radical generation is instantaneous, is preferred for kinetic work. The number of quanta absorbed by the system is estimated first by irradiating a uranyl oxalate solution in the reaction vessel. This reacts quantitatively and provides a measure of the number of radicals produced when the fraction of light absorbed is also known.

The initiator efficiency can be estimated by counting the number of chains formed, but this relies on knowledge of the termination mechanism and the absence of chain transfer reactions. A better technique is to make use of a radical scavenger such as DPPH, ferric chloride, or benzoquinone. As long as one is confident that a single molecule of inhibitor reacts with only one radical, a quantitative estimate of the number of radicals produced is possible. The addition of varying quantities of inhibitor to a system results in a corresponding number of induction periods whose durations are proportional to the number of radicals produced. This is illustrated in the photo-polymerization of vinyl acetate in the presence of benzoquinone.

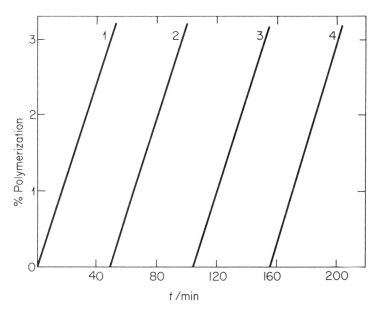

FIGURE 3.4. Inhibiting effect of benzoquinone on the photopolymerization of vinyl acetate (1) no quinone (2) 2.39 mg (3) 5.00 mg (4) 7.50 mg. (Adapted from Melville.)

Determination of k_p and k_t. To obtain individual values of k_p and k_t, a combination of two or more of the following measurements is required: (1) a steady-state measurement of v_p to give $(k_p^2 k_d/k_t)$; (2) an average radical life time at the steady state yielding $(k_d k_t)^{-1/2}$; (3) an estimate of x_n which gives $(k_p^2/k_t k_d)$.

MEASUREMENT OF v_p BY DILATOMETRY
The overall polymerization rate can be monitored through the change in a physical or chemical property of the system. Gravimetric analysis of the products or titration with bromine to measure the rate of disappearance of double bonds would both afford a means of following the reaction but present difficulties, especially if oxygen acts as an inhibitor. The choice of a physical property is more desirable. A change in the molar refractivity is related to the change in the number of double bonds, so that an increase in the refractive index could be used to follow v_p. Alternatively, one can make use of the fact that the density of the

polymer is normally greater than that of the monomer and volume contractions as large as 27 per cent of the total have been obtained during the course of a reaction. This makes dilatometry a particularly useful technique. If only low degrees of conversion are measured the initiator concentration may be assumed to be constant and if L_0 is the height of the initial level in the dilatometer, L_t that after time t, and L_∞ the reading on completion of the reaction, a plot of log $\{(L_0 - L_\infty)/(L_t - L_\infty)\}$ against t is linear if the reaction is first order in monomer, *i.e.* if $v_p = k_p(fk_d[I]/k_t)^{1/2}[M]$. The slope is then equal to the collection of constants and when k_d, f, and $[I]$ are known, $(k_p/k_t^{1/2})$ is obtained.

RADICAL LIFETIMES BY THE ROTATING SECTOR TECHNIQUE

Under steady-state conditions the average life-time of a propagating chain τ_s is determined by the ratio of the concentration of radicals at any time to their rate of disappearance, given by equation (3.11),

$$\tau_s = [M^\bullet]/2k_t[M^\bullet]^2 = 1/2k_t[M^\bullet]. \tag{3.21}$$

Substituting for the radical concentration (using equation (3.10)) gives

$$\tau_s = k_p[M]/2k_t v_p \tag{3.22}$$

and as $(k_p/k_t^{1/2})$ is obtained from v_p, knowledge of τ_s provides a means of separating k_p and k_t.

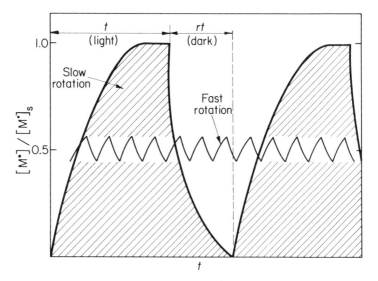

FIGURE 3.5. Variation of the radical concentration in a rotating sector photo-polymerization with time for slow and fast rotation speeds. This is expressed as the ratio of the radical concentration at time t, $[M^\bullet]$, to the concentration under steady-state conditions $[M^\bullet]_S$.

The radical life-time τ_s can be evaluated from non-steady-state conditions and this is conveniently carried out using a photo-initiated system where controlled instantaneous generation of radicals is possible. The reaction mixture is placed in a quartz cell dilatometer and thermostatted. The cell is then placed in a beam of u.v. radiation which can be interrupted to produce alternating periods of illumination and darkness, by rotating a disc with a cut-out sector in the light path. The ratio of the durations of the dark and light periods is r and a typical value is $r = 3$. The time of each period of illumination in a cycle can be varied by altering the rate of rotation of the disc and in this way both steady and non-steady-state conditions can be attained.

For slow speeds of rotation the illumination time t is large compared with τ_s, thereby allowing the radical concentration to build up to a steady state during each cycle (shaded areas in figure 3.5). If the cycle time is now increased until t is small compared with τ_s, then the radical concentration never reaches the steady state but remains low and almost constant. This in effect reduces the light intensity by a factor of $(1 + r)^{-1}$, and if ν_p is measured for various rotation times, the ratio (ν_p/ν_{ps}) will change from a lower limit of $(1 + r)^{-1}$ for large values of t to an upper limit of $(1 + r)^{-1/2}$ for small t, as the illumination time passes from $t > \tau_s$ to $t < \tau_s$. Here ν_p and ν_{ps} are the average and steady-state rates respectively. To locate the intermediate speed for which $t = \tau_s$ a plot of (ν_p/ν_{ps}) against log t is constructed and compared with a theoretical curve of (ν_p/ν_{ps}) against $(\log t - \log \tau_s)$. The two curves are superimposed by displacing the theoretical curve along the abscissa and the magnitude of the horizontal shift is then equal to $\log \tau_s$.

TABLE 3.3. Kinetic parameters for the photopolymerization of vinyl acetate

	Illumination	
	High intensity	Low intensity
ν_i/mol dm^{-3} s^{-1}	7.29×10^{-9}	1.11×10^{-9}
ν_{ps}/mol dm^{-3} s^{-1}	1.19×10^{-4}	0.45×10^{-4}
τ_s/s	1.50	4.00
$(k_p/k_t^{1/2})$/(dm^3 mol^{-1} s^{-1})$^{1/2}$	0.1826	0.177
(k_p/k_t)	3.3×10^{-5}	3.32×10^{-5}
k_p/dm^3 mol^{-1} s^{-1}	1.01×10^3	0.94×10^3
k_t/dm^3 mol^{-1} s^{-1}	3.06×10^7	2.83×10^7

KINETIC PARAMETERS
Some typical results for the photo-initiated polymerization of vinyl acetate at two intensities of radiation are given in table 3.3.

3.12 Activation energies and the effect of temperature
The influence of temperature on the course of a polymerization reaction depends on initiator efficiency and decomposition rate, chain transfer, and chain propagation, but it is important to have some knowledge of its effect in order to formulate optimum conditions for a reaction.

The energy of activation of a polymerization reaction is easily determined using an Arrhenius plot when the rate constants have been determined at several temperatures, but even for a simple reaction the overall rate is still a three-stage process and the total activation energy is the sum of the three appropriate contributions for initiation, propagation, and termination.

Remembering that ν_p is proportional to $k_p(k_d/k_t)^{1/2}$, the overall activation energy E_a is

$$E_a = \tfrac{1}{2}E_d + (E_p - \tfrac{1}{2}E_t). \tag{3.23}$$

The term $(E_p - \tfrac{1}{2}E_t)$ provides a measure of the energy required to polymerize a particular monomer and has been estimated for styrene to be 27.2 kJ mol^{-1} and for vinyl acetate 19.7 kJ mol^{-1}. Initiators have values of E_d in the range 125 to 170 kJ mol^{-1} and this highlights the controlling role of the initiation step in free radical polymerization. Consequently values of E_a are generally in the range 85 to 150 kJ mol^{-1}.

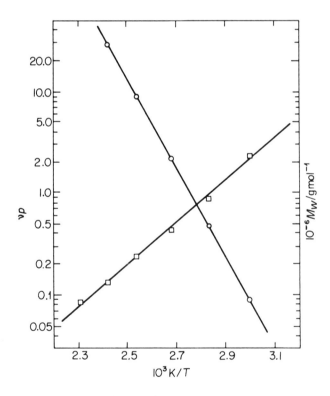

FIGURE 3.6. Dependence of the polymerization rate ν_p, expressed as per cent conversion per hour, ╋ o ╋ and polymer molar mass ╋ □ ╋ on temperature, for the thermal, self-initiated polymerization of styrene.

TABLE 3.4. Parameters in typical radical chain polymerizations

Monomer	$\dfrac{10^{-3} k_p}{\text{dm}^3 \text{ mol}^{-1} \text{ s}^{-1}}$	$\dfrac{E_p}{\text{kJ mol}^{-1}}$	$\dfrac{A_p}{\text{dm}^3 \text{ mol}^{-1} \text{ s}^{-1}}$	$\dfrac{10^{-7} k_t}{\text{dm}^3 \text{ mol}^{-1} \text{ s}^{-1}}$	$\dfrac{E_t}{\text{kJ mol}^{-1}}$	$\dfrac{10^{-9} A_t}{\text{dm}^3 \text{ mol}^{-1} \text{ s}^{-1}}$
Vinyl chloride	12.3	15.5	0.33	2300	17.6	600
Acrylonitrile	1.96	16.3	—	78.2	15.5	—
Methyl acrylate	2.09	29.7	10	0.95	22.2	15
Methyl methacrylate	0.705	19.7	0.087	2.55	5.0	0.11
1,3-Butadiene	0.100	38.9	12	—	—	—

† Here A is the collision frequency factor in the Arrhenius equation $k = A \exp(-E/RT)$.

Typical values for these quantities are shown in table 3.4. Since the temperature term in the rate equation is $\exp\{(\frac{1}{2}E_t - \frac{1}{2}E_d - E_p)/RT\}$, the exponent will normally be negative so that the polymerization rate will increase as the temperature is raised. The change in molar mass can also be examined in this way and now the quantity $\{k_p/(k_d k_t)^{1/2}\}$ is the one of interest. The required energy term is $\exp\{(E_p - \frac{1}{2}E_d - \frac{1}{2}E_t)/RT\}$ and in thermal polymerizations this is negative and usually about -60 kJ mol^{-1}. As the temperature increases, the chain length decreases rapidly and only in pure photochemical reactions, where E_d is zero, is the activation energy slightly positive, leading to a modest increase in x_n as the temperature goes up.

3.13 Thermodynamics of radical polymerization

The conversion of an alkene to a long chain polymer has a negative enthalpy (ΔH_p is negative) because the formation of a σ-bond from a π-bond is an exothermic process. While the enthalpy change favours the polymerization, the change in entropy is unfavourable and negative because the monomer becomes incorporated into a covalently bonded chain structure. However, examination of the relative magnitudes of the two effects shows that whereas $-\Delta S_p$ is in the range 100 to 130 J K^{-1} mol^{-1}, $-\Delta H_p$ is normally 30 to 150 kJ mol^{-1}. The overall Gibbs free energy change $\Delta G_p = \Delta H_p - T\Delta S_p$ is then negative and the polymerization is thermodynamically feasible.

These conditions favour the formation of polymer, but it is obvious from the general treatment of the energetics of the reaction that the chain length decreases as the temperature rises. This can be understood if we postulate the existence of a depolymerization reaction.

When the temperature increases, the depolymerization reaction becomes more important and ΔG_p becomes less negative. Eventually a temperature is reached at which $\Delta G_p = 0$ and the overall rate of polymerization is zero. This temperature is known as the *ceiling temperature* T_c.

If both the forward and reverse reactions are treated as chain reactions then

$$M_n^{\bullet} + M \underset{k_{dp}}{\overset{k_{p_\bullet}}{\rightleftharpoons}} \sim\!\!M_{n+1}^{\bullet}$$

where k_{dp} is the rate constant for the depropagation. The overall rate expression is then obtained by modifying equation (3.10),

$$v_p = k_p [M^{\bullet}] [M] - k_{dp} [M^{\bullet}] \qquad (3.24)$$

while the degree of polymerization x is

$$x = (k_p [M] - k_{dp} [M^{\bullet}])/r_t. \qquad (3.25)$$

At the ceiling temperature $v_p = 0$ and so

$$k_p [M] = k_{dp}. \qquad (3.26)$$

This illustrates that the ceiling temperature is a specific value for a given concentration [M] of monomer.

The influence of temperature on the polymerization of styrene is shown in figure 3.7. The factor k_p[M] increases steadily from 300 K, whereas k_{dp} only becomes significant around 460 K but increases more rapidly than k_p[M] until the two curves intersect at T_c. This also illustrates that no polymer can be formed above T_c from monomer at concentration [M]. Some values for T_c of pure monomers are shown in table 3.5.

3.14 Features of free radical polymerization

The main features of a radical polymerization can now be summarized and contrasted with the corresponding step-growth reactions. (c.f. page 35)

(1) A high molar mass polymer is formed immediately the reaction begins and the average chain length shows little variation throughout the course of the polymerization.

(2) The monomer concentration decreases steadily throughout the reaction.

(3) Only the active centre can react with monomer and add units onto the chain one after the other.

(4) Long reaction times increase the polymer yield, but not the molar mass of the polymer.

(5) An increase in temperature increases the rate of the reaction but decreases the molar mass.

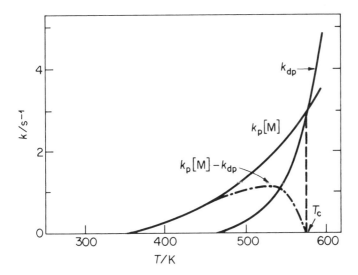

FIGURE 3.7. The temperature dependence of k_p[M] and k_{dp} for styrene. (After Dainton and Ivin.)

TABLE 3.5. Ceiling temperatures T_c for pure monomers

Monomer	T_c/K
Styrene	583
Methyl methacrylate	493
α-Methyl styrene	334

3.15 Polymerization processes

Industrial radical initiated polymerizations can be carried out in one of four different ways:
(a) with monomer only — *bulk;*
(b) in a solvent — *solution;*
(c) with monomer dispersed in an aqueous phase — *suspension;*
(d) or as an *emulsion.*

Bulk polymerization is used in the production of polystyrene, poly(methyl methacrylate), and poly(vinyl chloride). The reaction mixture contains only monomer and initiator, but because the reaction is exothermic, hot spots tend to develop when heat removal is inefficient. Auto-acceleration occurs in the highly viscous medium making control difficult and efficient monomer conversion is impeded. To overcome some of the disadvantages low conversions and used, after which the unreacted monomer is stripped off and recycled. The main advantages of the technique lie in the optical clarity of the product and its freedom from contaminations.

Quiescent mass polymerization is an unstirred reaction used for casting sheets of poly(methyl methacrylate). A low molar mass prepolymer is prepared and then the main polymerization is carried out *in situ* making use of the Tromsdorff effect to obtain high molar mass material and tougher sheets. The two stage approach helps to control the heat evolved.

In *solution polymerization,* the presence of the solvent facilitates heat transfer and reduces the viscosity of the medium. Unfortunately the additional complication of chain transfer arises and solvents must be selected with care.

Ethylene, vinyl acetate, and acrylonitrile are polymerized in this way. The redox initiated polymerization of acrylonitrile is an example of precipitation polymerization where the polyacrylonitrile formed is insoluble in water and separates as a powder. This can lead to undesirable side reactions known as popcorn polymerizations when tough crosslinked nodules of polymer grow rapidly and foul the feed lines in industrial plants.

Suspension polymerization counteracts the heat problem by suspending droplets of water-insoluble monomer in an aqueous phase. The droplets are obtained by vigorous agitation of the system and are in the size range 0.01 to 0.5 cm diameter. The method is in effect a bulk polymerization which avoids the complications of heat and viscosity build-up.

Emulsion polymerization is an important technological process widely used to prepare acrylic polymers, poly(vinyl chloride), poly(vinyl acetate), and a large number of copolymers. The technique differs from the suspension method in that the particles in the system are much smaller, 0.05 to 5 μm diameter, and the initiator is soluble in the aqueous phase rather than in the monomer droplets. The process offers the unique opportunity of being able to increase the polymer chain length without altering the reaction rate. This can be achieved by changing either the temperature or the initiator concentration, and the reasons for this will become more obvious when we examine the technique more closely.

The essential ingredients are monomer, emulsifying agent, water, and a water-soluble initiator. The surfactant is normally an amphipathic long chain fatty acid salt with a hydrophilic "head" and a hydrophobic "tail". In aqueous solutions these form aggregates or micelles (0.1 to 0.3 μm long), consisting of 50 to 100 molecules oriented with the tails inwards, thereby creating an interior hydrocarbon environment and a hydrophilic surface of heads in contact with

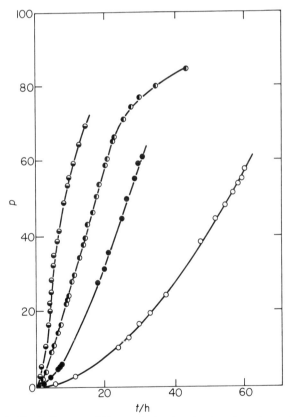

FIGURE 3.8. The influence of the surfactant potassium laurate on the emulsion polymerization of isoprene at 323 K. The percentage polymerization p is shown as a function of time t for four concentration of potassium laurate: ∘, 0.01 mol dm^{-3}; •, 0.04 mol dm^{-3}; ◑, 0.10 mol dm^{-3}; and ◐, 0.50 mol dm^{-3}.

the water. The micelles exist in equilibrium with free molecules in the aqueous
phase and the concentration must exceed the "critical micelle concentration"
of the emulsifier.

When monomer is added to the dispersion, the bulk of it remains in the
aqueous phase as droplets but some dissolves in the micelles, swelling them. Free
radicals are generated from a water-soluble redox system such as persulphate +
ferrous

$$S_2O_8^{2-} + Fe^{2+} \rightarrow Fe^{3+} + SO_4^{2-} + SO_4^{-\bullet}$$

at a rate of 10^{16} dm^{-3} s^{-1}. The radicals diffuse through the aqueous phase and
penetrate both the micelles and droplets but as the concentration of micelles
(about 10^{21} dm^{-3}) far exceeds that of the droplets (10^{13} to 10^{14} dm^{-3}), polymerization is centred almost exclusively in the micelle interior. After only 2 to
10 per cent conversion the character of the system has changed markedly.
Constant replenishment of the polymer swollen micelles takes place by diffusion
from the droplets which decrease steadily in size until at about 50 to 80
per cent conversion they have been totally consumed. Polymerization then
continues at a steadily decreasing rate until all the remaining monomer in the
micelle is converted into polymer.

The polymerization is a radical initiated process and the propagation rate
v_p is proportional to the concentration of monomer in the swollen micelle [M] and
the radical concentration $[M^\bullet]$.

$$v_p = k_p [M] [M^\bullet] \qquad (3.27)$$

The analysis assumes that only one radical can be tolerated in a micelle at any
one time and the polymerization proceeds until a second radical diffuses in
to terminate the reaction. The process begins again when another radical
penetrates the micelle. Consequently the polymerization in any micelle is only
active for half the time and so $[M^\bullet] = N/2$ where N is the total number of
activated micelles per unit volume of emulsion. In the absence of chain transfer,
x_n is given by

$$x_n = k_p N [M] / 2v_i , \qquad (3.28)$$

for a coupling termination reaction. As the number of micelles is proportional
to the surfactant concentration, both v_p and x_n are also functions of this
parameter, and this is illustrated in figure 3.8 for the polymerization of isoprene
at 323 K using four concentrations of potassium laurate. Thus a combination of
high rates and large x_n can be obtained without temperature variation and
this provides the system with its particular appeal. Control of the chain length
can be achieved, when desired, by adding a chain transfer agent such as dodecyl
mercaptan.

3.16 Emulsion polymerization of SBR
Natural rubber is still one of the best elastomers available, but it is in competition

with a number of excellent synthetic substitutes. One of these, the styrene-butadiene copolymer SBR, is now the most widely used elastomer.

SBR can be prepared using a free radical emulsion process, which produces a random copolymer containing 25 per cent of styrene and 75 per cent of butadiene and having T_g = 213 K. A typical recipe for this synthesis, at a temperature of 323 K and a pressure of 1.73×10^6 N m^{-2} is

Butadiene	75 parts by weight
Styrene	25 parts by weight
Water	180 parts by weight
Soap	5 parts by weight
Dodecyl mercaptan	0.5 parts by weight
Potassium persulphate	0.5 parts by weight

This produces the so-called "hot" rubber, a random copolymer with 60 per cent of the butadiene content in the *trans*-1,4 configuration, about 20 per cent in the *cis*-1,4 form and 20 per cent in the 1,2 configuration. Alternatively, the temperature of polymerization can be lowered to 278 K and initiated using a redox couple. This produces the "cold" rubber, with an increased *trans*-1,4 content, which is more popular commercially.

General Reading
H. Alter and A. D. Jenkins, "Chain-reaction polymerization" in *Encyclopaedia of Polymer Science and Technology*. Interscience Publishers Inc. (1965).
C. E. H. Bawn, *The Chemistry of High Polymers*. Butterworths (1948).
F. W. Billmeyer, *Textbook of Polymer Science*. John Wiley and Sons, 1962.
E. W. Duck, "Emulsion polymerization" in *Encyclopaedia of Polymer Science and Technology*. Interscience Publishers Inc. (1966).
P. J. Flory, *Principles of Polymer Chemistry*, Chapter 4. Cornell University Press, Ithaca, N.Y. (1953).
G. E. Ham, *Vinyl Polymerization*, Vol. I. Marcel Dekker (1967).
A. D. Jenkins, "The reactivity of polymer radicals" in *Adv. in Free Radical Chemistry*, Vol. 2. Logos Press Ltd. (1967).
R. W. Lenz, *Organic Chemistry of Synthetic High Polymers*, Chapters 9–11. Interscience (1967).
D. Margerison and G. C. East, *Introduction to Polymer Chemistry*, Chapter 4. Pergamon Press (1967).
G. Odian, *Principles of Polymerization*, Chapter 3. McGraw-Hill (1970).
S. R. Palit, S. R. Chatterjee, A. R. Mukherjee, "Chain transfer" in *Encyclopaedia of Polymer Science and Technology*, Vol. 3. Interscience Publishers Inc. (1965).
H. Ringsdorf, "Bulk polymerization" in *Encyclopaedia of Polymer Science and Technology*, Interscience Publishers Inc. (1965).
R. B. Seymour, "Polymerization techniques", *Plastics World*, **25**, 54 (1967).
D. A. Smith, *Addition Polymers*, Chapter 2. Butterworths (1968).

References
1. F. S. Dainton and K. J. Ivin, *Quarterly Reviews*, **12**, 61 (1958).
2. H. W. Melville, *J. Chem. Soc.*, 247 (1947).
3. G. V. Schulz, *Ber.*, **80**, 232 (1947).

CHAPTER 4

Ionic Polymerization

4.1 General characteristics

Radical-initiated polymerizations are generally non-specific, but this is not true for ionic initiators, since the formation and stabilization of a carbonium ion or carbanion depends largely on the nature of the group R in the vinyl monomer CH_2=CHR. For this reason cationic initiation is usually limited to monomers with electron-donating groups which help to stabilize the delocalization of the positive charge in the π-orbitals of the double bond. Anionic initiators require electron withdrawing substituents (—CN, —COOH, —CH=CH$_2$, *etc.*) to promote the formation of a stable carbanion, and when there is a combination of both mesomeric and inductive effects the stability is greatly enhanced.

As these ions are associated with a counter-ion or gegen-ion the solvent has a profound influence. Chain propagation will depend significantly on the separation of the two ions and this separation will also control the mode of entry of an adding monomer. Also, the gegen-ion itself can influence both the rate and stereochemical course of the reaction. While polar and highly solvating media are obvious choices for ionic polymerizations, many cannot be used because they react with and negate the ionic initiators. This is true of the hydroxyl solvents, and even ketones will form stable complexes with the initiator to the detriment of the reaction. As solvents of much lower dielectric constant have to be used, the resulting close proximity of the gegen-ion to the chain end requires that one must treat the propagating species as an *ion pair,* but even in low polarity media such as methylene chloride, ether, THF, nitrobenzene, *etc.*, the ion pair separation can vary sufficiently for the effects to be distinguishable.

Ionic-initiated polymerizations are much more complex than radical reactions. When the chain carrier is ionic, the reaction rates are rapid, difficult to reproduce, and yield high molar mass material at low temperatures by mechanisms which are often difficult to define.

Complications in the kinetic analysis can arise from co-catalyst effects where small quantities of an inorganic compound, such as water, will have an unexpectedly large influence on the polymerization rate.

Initiation of an ionic polymerization can occur in one of four ways involving essentially the loss or gain of an electron e^- by the monomer to produce an ion or radical ion.

(a) $M + I^+ \rightarrow MI^+$ Cationic
(b) $M + I^- \rightarrow MI^-$ Anionic
(c) $M + e^- \rightarrow \,^{\bullet}M^-$ Anionic
(d) $M - e^- \rightarrow \,^{\bullet}M^+$ Cationic (charge transfer)

4.2 Cationic polymerization

Ionic polymerizations proceed by a chain mechanism and can be dealt with under the general headings that were used for the radical reactions: initiation, propagation, and termination. A common type of cationic initiation reaction is that represented in (a) where I^+ is typically a strong Lewis acid. These electrophilic initiators are classed in three groups: (1) classical protonic acids or acid surfaces – HCl, H_2SO_4, $HClO_4$; (2) Lewis acids or Friedel-Crafts catalysts – BF_3, $AlCl_3$, $TiCl_4$, $SnCl_4$; (3) carbonium ion salts.

The most important initiators are the Lewis acids MX_n, but they are not particularly active alone and require a co-catalyst SH to act as a proton donor. In general

$$MX_n + SH \rightleftharpoons [SMX_n]^- H^+$$

and a probable initiation mechanism is the two-step process.

Where step one is the rapid formation of a π-complex and step two is a slow intramolecular rearrangement. While the need for a co-catalyst is recognized it is often difficult to demonstrate, and a useful reaction which serves this purpose is the polymerization of isobutylene. This reaction proceeds rapidly when trace quantities of water are present but remains dormant under anhydrous conditions.

The active catalyst — co-catalyst species required to promote this reaction is

$$BF_3 + H_2O \rightleftharpoons H^+[BF_3OH]^-,$$

and the complex reacts with the monomer to produce a carbonium ion chain carrier which exists as an ion pair with $[BF_3OH]^-$.

$$H^+[BF_3OH]^- + (CH_3)_2C{=}CH_2 \rightarrow (CH_3)_3C^+[BF_3OH]^-$$

The type of co-catalyst also influences the polymerization rate because the activity of the initiator complex depends on how readily it can transfer a proton to the monomer. If the polymerization of isobutylene is initiated by $SnCl_4$, the acid strength of the co-catalyst governs the rate, which decreases in the co-catalyst order: acetic acid > nitroethane > phenol > water.

Other types of initiator are less important; thus strong acids protonate the double bond of a vinyl monomer

$$HA + CH_2{=}CR_1R_2 \rightarrow A^-CH_3C^+R_1R_2,$$

while iodine initiates polymerization with the ion pair

$$2I_2 \rightarrow I^+I_3^-$$

which forms a stable π-complex with olefins such as styrene and vinyl ethers.

$$CH_2{=}CHR + I^+I_3^- \rightarrow ICH_2{-}\overset{+}{C}HI_3^-$$
$$\underset{R}{|}$$

A recent suggestion that it may be a charge transfer mechanism has not been fully substantiated.

High energy radiation is also thought to produce cationic initiation, but this may lead to fragmentation and a mixture of free radical and cationic centres.

4.3 Propagation by cationic chain carriers

Chain growth takes place through the repeated addition of a monomer in a head-to-tail manner, to the carbonium ion, with retention of the ionic character throughout.

$$CH_3{-}\overset{R_1}{\underset{R_2}{\overset{|}{\underset{|}{C^+}}}}[SMX_n]^- + nCH_2{=}CR_1R_2 \overset{k_p}{\rightarrow} CH_3[CR_1R_2CH_2]_n \overset{R_1}{\underset{R_2}{\overset{|}{\underset{|}{C^+}}}}[SMX_n]^-$$

The mechanism depends on the *counterion*, the *solvent*, the *temperature*, and the *type of monomer*. Reactions can be extremely rapid when strong acid initiators such as BF_3 are used, and produce long chain polymer at low temperatures. Rates tend to be slower when the weaker acid initiators are used and a polymerization with $SnCl_4$ may take several days. Useful reaction temperatures are in the range 170 to 190 K and both molar mass and reaction rate decrease as the temperature is raised.

Propagation also depends greatly on the position and type of the gegen-ion associated with the chain carrier. The position of the gegen-ion can be altered by varying the dielectric constant of the solvent and large changes in k_p can be obtained as shown in table 4.1 for a perchloric acid initiated polymerization of styrene in several media.

TABLE 4.1. Cationic polymerization of styrene in media of varying dielectric constant ϵ

Solvent	ϵ	Catalyst	$\dfrac{k_p}{dm^3 \ mol^{-1} \ s^{-1}}$
CCl_4	2.3	$HClO_4$	0.0012
$CCl_4 + (CH_2Cl)_2$ (40/60)	5.16	$HClO_4$	0.40
$CCl_4 + (CH_2Cl)_2$ (20/80)	7.0	$HClO_4$	3.20
$(CH_2Cl)_2$	9.72	$HClO_4$	17.0
$(CH_2Cl)_2$	9.72	$TiCl_4/H_2O$	6.0
$(CH_2Cl)_2$	9.72	I_2	0.003

It has been suggested that the various stages of the ionization producing carbonium ions can be represented as

$$ RX \quad \rightleftharpoons \quad R^+X^- \quad \rightleftharpoons \quad R^+//X^- \quad \rightleftharpoons \quad R^+ + X^-. $$

covalent   intimate ion pair   solvent separated   free ions
                                              ion pair

The increasing polarity of the solvent alters the distance between the ions from an intimate pair, through a solvent separated pair to a state of complete dissociation. As free ions propagate faster than a tight ion pair, the increase in free ion concentration with change in dielectric constant is reflected in an increase in k_p. The separation of the ions also lowers the steric restrictions to the incoming monomer, so that free ions exert little stereo-regulation on the propagation and too great a separation may even hinder reactions which are assisted by co-ordination of the monomer with the metal in the gegen-ion. To the first approximation it can be stated that as the dielectric constant of the medium increases, there is a linear increase in the polymer chain length and an exponential increase in the reaction rate, but in some cases the bulk dielectric of the medium may not determine the effect of the solvent on an ion in its immediate environment. This leads to deviations from the simple picture. The nature of the gegen-ion affects the polymerization rate. Larger and less tightly bound ions lead to larger values of k_p, hence a decrease in k_p is observed as the initiator changes from $HClO_4$ to $TiCl_4 . H_2O$ to I_2, for the reaction of styrene in 1,2-dichloro-ethane.

4.4 Termination

The termination reaction in a cationic polymerization is less well defined than for the radical reactions, but is thought to take place either by a unimolecular rearrangement of the ion pair

$$\sim\!\!CH_2\overset{\displaystyle R_1}{\underset{\displaystyle R_2}{\overset{\mid}{\underset{\mid}{C}}}}{}^+\,[SMX_n]^- \rightarrow \sim\!\!CH\!=\!CR_1R_2 + H^+\,[SMX_n]^-,$$

or through a bimolecular transfer reaction with a monomer

$$\sim\!\!CH_2\overset{\displaystyle R}{\underset{\displaystyle R_2}{\overset{\mid}{\underset{\mid}{C}}}}{}^+\,[SMX_n]^- + CH_2\!=\!\overset{\displaystyle R_1}{\underset{\displaystyle R_2}{\overset{\mid}{\underset{\mid}{C}}}} \rightarrow \sim\!\!CH\!=\!CR_1R_2 + CH_3\overset{\displaystyle R_1}{\underset{\displaystyle R_2}{\overset{\mid}{\underset{\mid}{C}}}}{}^+\,[SMX_n]^-$$

The first involves hydrogen abstraction from the growing chain to regenerate the catalyst-co-catalyst complex, while the second re-forms a monomer-initiator complex, thereby ensuring that the kinetic chain is not terminated by the reaction. In the unimolecular process, actual covalent combination of the active centre with a catalyst-co-catalyst complex fragment may occur giving two inactive species. This serves to terminate the kinetic chain and reduce the initiator complex and, as such, is a more effective route to reaction termination.

4.5 General kinetic scheme

Many cationic polymerizations are both rapid and heterogeneous which makes the formulation of a rigorous kinetic scheme extremely difficult. At best, one of general validity can be deduced, but this should not be applied indiscriminately. Following the steady-state approach outlined for radical reactions, the rate of initiation v_i of a cationic reaction is proportional to the catalyst-co-catalyst concentration c and the monomer concentration $[M]$.

$$v_i = k_i c\,[M] \tag{4.1}$$

Termination can be taken as a first-order process in contrast to the free radical mechanisms, and

$$v_t = k_t\,[M^+] \tag{4.2}$$

Under steady-state conditions, $v_i = v_t$ and

$$[M^+] = k_i c\,[M]\,/k_t. \tag{4.3}$$

This gives an overall polymerization rate v_p of

$$v_p = k_p[M][M^+] = (k_p k_i/k_t)c[M]^2 \qquad (4.4)$$

and a chain length of

$$x_n = v_p/v_t = (k_p/k_t)[M], \qquad (4.5)$$

if termination, rather than transfer, is the dominant process. When chain transfer is significant,

$$x_n = k_p/k_{tr}. \qquad (4.6)$$

Although not universally applicable, this scheme gives an adequate description of the polymerization of styrene by $SnCl_4$ in ethylene dichloride at 298 K.

4.6 Energetics of cationic polymerization

Having established a kinetic scheme, some explanation for the increase in overall rate with decreasing temperature may be forthcoming. The rate is proportional to $(k_i k_p/k_t)$ so the overall activation energy E is given by

$$E = E_i + E_p - E_t, \qquad (4.7)$$

and for the chain length

$$E_x = E_p - E_t. \qquad (4.8)$$

Propagation in a cationic polymerization requires the approach of an ion to an uncharged molecule in a relatively non-polar medium and as this is an operation with a low activation energy, E_p is much less than E_i, E_t, or E_{tr}. Consequently E is normally in the range -40 to $+60$ kJ mol^{-1}, and when it is negative, the rather unexpected increase in k_p is obtained with decreasing temperature. It should be noted, however, that not all cationic polymerizations have negative activation energies; the polymerizations of styrene by trichloroacetic acid in nitromethane and by 1,2-dichloroethylene have E equal to $+57.8$ kJ mol^{-1} and $+33.6$ kJ mol^{-1} respectively.

The chain length, on the other hand, will always decrease as the reaction temperature rises because E_t is always $> E_p$.

4.7 Experimental cationic polymerization

A typical cationic reaction can be demonstrated by the polymerization of trioxan to form polyformaldehyde and using boron trifluoride etherate as initiator. The polymer formed can then be stabilized by acetylating the terminal hydroxyl group

with acetic anhydride.

The catalyst solution of $BF_3O(C_2H_5)_2$ (2 cm^3) in benzene (40 cm^3) is prepared and thoroughly dried over calcium hydride before starting the polymerization. A mixture of cyclohexane (12 cm^3), previously dried over calcium hydride, and 10.0 g of trioxan, recrystallized from methylene chloride, is added to a dry test tube. The tube is loosely stoppered and placed in a thermostat bath at 340 K. The catalyst solution (0.15 cm^3) is then added to the reaction mixture using a syringe, and the tube recapped. The reaction proceeds rapidly and polymer forms in about 5 min, after which time the bath temperature is lowered to 320 K for about 30 min. The product is separated by filtration and washed with isopropanol containing 10 per cent aqueous ammonia, then dried in *in vacuo* at 330 K overnight. The polyformaldehyde is soluble in dimethyl formamide.

4.8 Stable carbonium ions
Many of the uncertainties inherent in Friedel-Crafts catalyst-co-catalyst systems can be removed if stable, well defined, initiators are used. Bawn and his co-workers have made use of triphenyl methyl and tropylium salts of the general formula $Ph_3C^+X^-$ and $C_7H_7^+X^-$ where X^- is a stable anion such as ClO_4^-, $SbCl_6^-$, and PF_6^-.

Initiation occurs by one of three mechanisms:

(i) Direct addition: $I^+ + CH_2{=}CHR \rightarrow ICH_2{-}\overset{+}{C}HR$

(ii) Hydride extraction: $I^+ + CH_2{=}CHR \rightarrow IH + CH{\cdots}\overset{+}{C}H{-}R$

(iii) Electron transfer: $I^+ + CH_2{=}CHR \rightarrow I^{\cdot} + {}^{\cdot}[CH_2{=}CHR]^+$

The reaction of trityl hexafluorophosphate and tetrahydrofuran (THF) has been shown to proceed without evidence of a termination reaction and a "living" cationic system can be obtained. The reaction takes place below room temperature.

The effect of the counter-ion is a noticeable factor in the elimination of the termination reaction and neither $SbCl_6^-$ nor any other anion studied has proved as good as PF_6^-.

Other similar non-terminating systems have been identified, but the influence of the anion on the efficiency of the system to produce "living" polymers varies from monomer to monomer.

4.9 Cationic charge transfer initiation
Initiation of polymerization by mechanism (d) of section 4.1 requires that the monomer concerned is a nucleophile with respect to the electron acceptor which may be a halogen, a metal cation in a high oxidation state, nitrogen peroxide,

sulphur dioxide, or an organic molecule with an electrophilic group. The extent of the interaction between the donor D and the acceptor A determines the relative magnitudes of the constants in the equilibria

$$D + A \overset{k_1}{\rightleftharpoons} [D \ldots A \leftrightarrow D^{\delta+} \ldots A^{\delta-}] \overset{k_2}{\rightleftharpoons} [D^+ A^-] \overset{k_3}{\rightleftharpoons} \,^\cdot D^+ + \,^\cdot A^-,$$

but the condition $k_1 \gg k_2$, k_3 usually predominates, except for powerful donor-acceptor pairs. When the monomer is the donor its character determines the most propitious acceptor to initiate the reaction, thus N-vinyl carbazole can be polymerized in the presence of relatively weak acceptors such as sulphur dioxide, tetrachloroethylene, or acrylonitrile, where the intense colours usually associated with charge-transfer complexes rarely develop. Medium to strong acceptors, o-chloroanil, tetranitromethane, etc., can also be used. Ellinger has suggested the following scheme with acrylonitrile as the electrophile and where all charges are actually partial charges.

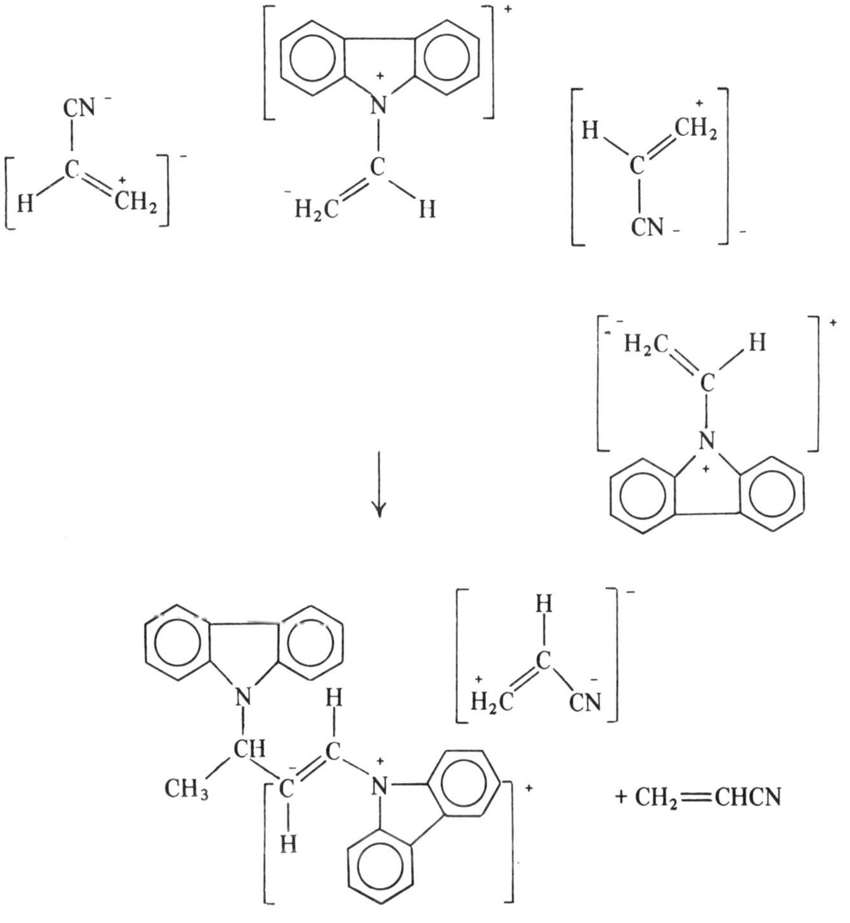

Vinyl ethers are different because they generally require a much more positive reaction with complete electron transfer to a strong acceptor, *e.g.* P_2O_5, triethanolamine, or tetracyanoethylene.

Some initiators may lead to propagation by both radical and cationic mechanisms. Most notable among these are the tropylium salts, studied by Bawn and Ledwith, who have suggested that the salt forms a charge-transfer complex with the monomer. This eventually dissociates to produce a tropyl radical and a monomer cation radical where X^- may be $SbCl_6^-$ or BF_4^-.

$$\bigcirc{+}\rangle X^- + CH_2{=}CHOR \rightleftharpoons (\text{complex}) \rightarrow \bigcirc\rangle \cdot + {}^{\cdot}[CH_2{-}CH{=}\overset{+}{OR}]X^-$$

The cation radical can dimerize or form a di-cation to propagate the chain, although this has not yet been proved conclusively. However, it has been shown that the polymerizations initiated by the tropylium salts are undoubtedly cationic in nature.

4.10 Anionic polymerization

The polymerization of monomers with strong electronegative groups – acrylonitrile, vinyl chloride, styrene, and methyl methacrylate – can be initiated by either mechanism (b) or (c) of section 4.1.

In (b) an ionic or ionogenic molecule is required, capable of adding the anion to the vinyl double bond and so create a carbanion.

$$CX \rightarrow C^+ + X^-$$
$$X^- + M \rightarrow MX^-$$

The gegen-ion C^+ may be inorganic or organic and typical initiators include KNH_2, *n*-butyl lithium, and Grignard reagents (alkyl magnesium bromides).

If the monomer has a strong electron withdrawing group, then only a weakly positive initiator (Grignard) will be required for polymerization, but when the side group is phenyl or the electronegativity is low, a highly electropositive metal initiator, such as a lithium compound, is needed.

Mechanism (c) is the direct transfer of an electron from a donor to the monomer to form a radical anion. This can be accomplished by means of an alkali metal, and Na or K can initiate the polymerization of butadiene and methacrylonitrile; the latter reaction is carried out in liquid ammonia at 198 K.

$$Na + CH_2{=}\underset{\underset{CN}{|}}{\overset{\overset{CH_3}{|}}{C}} \rightarrow Na^+ + {}^{\cdot}\left[CH_2{-}\underset{\underset{CN}{|}}{\overset{\overset{CH_3}{|}}{C^-}}\right]$$

The anionic reactions have characteristics similar in many ways to the cationic polymerizations. In general they are rapid at low temperatures but are slower and less sensitive to changes in temperature than the cationic reactions. Reaction rates depend on the dielectric constant of the solvent, the resonance stability of the carbanion, the electronegativity of the initiator, and the degree of solvation of the gegen-ion. Many anionic polymerizations have no formal termination step but are sensitive to traces of impurities and as carbanions are quickly neutralized by small quantities of water, alcohol, carbon dioxide, and oxygen these are effective terminating agents. This imposes the need for rather rigorous experimental procedures to exclude impurities when anionic polymerizations are being studied and a few of these procedures will be mentioned later.

4.11 Polymerization of styrene by KNH_2

One of the first anionic reactions studied in detail was the polymerization of styrene in liquid ammonia, with potassium amide as initiator, reported by Higginson and Wooding. This serves to illustrate the general mechanism encountered and has the added interest that it is one of the few reactions involving free ions rather than ion pairs. Polymerizations were performed at 240 K in a highly polar medium, liquid ammonia.

Initiation is a two-step process; the dissociation of the potassium amide, first into its constituent ions, followed by addition of the anion to the monomer to create an active chain carrier.

$$KNH_2 \rightleftharpoons K^+ + :NH_2^-$$

$$:NH_2^- + CH_2{=}CHC_6H_5 \xrightarrow{k_i} H_2NCH_2{-}\overset{..}{\underset{..}{C}}HC_6H_5$$

The second step is rate determining so that

$$v_i = k_i c\,[M]\,, \tag{4.9}$$

where c is the concentration of the ion $:NH_2^-$.

Propagation is then the usual addition of monomer to the carbanion and the rate is given by

$$v_p = k_p[M]\,[M^-]\,. \tag{4.10}$$

Termination of the growing chain occurs when there is transfer to the solvent with regeneration of the amide ion, which is usually capable of initiating another chain.

$$H_2N{-}(CH_2CH)_n\,CH_2{-}\overset{\overset{\displaystyle H}{|}}{\underset{\underset{\displaystyle Ph}{|}}{C^-}} + NH_3 \xrightarrow{k_{tr}} NH_2{-}(CH_2{-}CHPh)_n\,CH_2CH_2Ph + :NH_2^-$$

The rate of termination is then

$$v_t = k_{tr}\,[M^-]\,[NH_3]\,. \tag{4.11}$$

The assumption of steady-state conditions gives an expression for the concentration of propagating polycarbanions,

$$[M^-] = (k_i/k_{tr})c[M]/[NH_3],$$ (4.12)

giving

$$v_p = (k_p k_i/k_{tr})c[M]^2/[NH_3]$$ (4.13)

and

$$x_n = (k_p/k_{tr})[M]/[NH_3].$$ (4.14)

The activation energy for the transfer process is larger than that for propagation and so the chain length decreases with increasing temperature, but as the overall activation energy for the reaction is positive, $+38$ kJ mol^{-1}, the reaction rate decreases with decreasing temperature.

4.12 "Living" polymers

The reaction scheme proposed for the initiation with potassium amide contains no formal termination step and if all the impurities which are liable to react with the carbanions are excluded from the system, propagation should continue until all monomer has been consumed, leaving the carbanion intact and still active. This means that if more monomer could be introduced, the active end would continue growing unless inadvertently terminated. These active polycarbanions were first referred to as "living" polymers by Szwarc.

One of the first "living" polymer systems studied was the polymerization of styrene initiated by sodium naphthalene. The initiator is formed by adding sodium to a solution of naphthalene in an inert solvent, tetrahydrofuran.

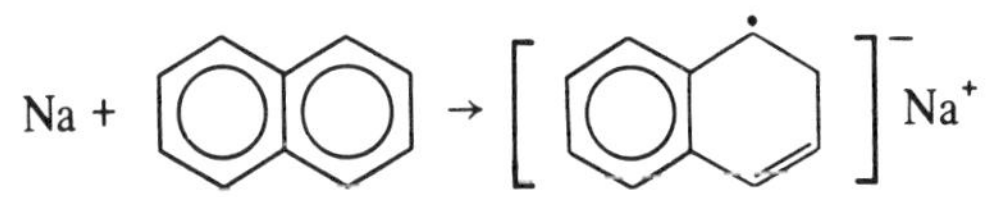

$$Na + \text{(naphthalene)} \rightarrow [\text{(naphthalene radical anion)}]^{-} Na^{+}$$

The sodium dissolves to form an addition compound and, by transferring an electron, produces the green naphthalene anion radical. Addition of styrene to the system leads to electron transfer from the naphthyl radical to the monomer to form a red styryl radical anion.

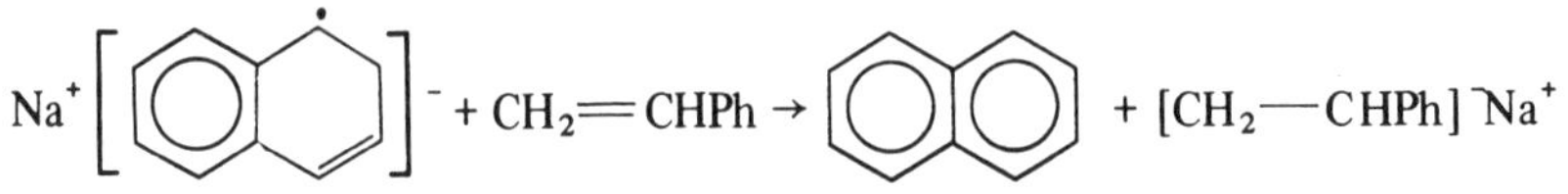

$$Na^{+}[\text{(naphthalene radical anion)}]^{-} + CH_2{=}CHPh \rightarrow \text{(naphthalene)} + [CH_2{-}CHPh]^{-} Na^{+}$$

It is thought, a dianion is finally formed capable of propagating from both ends.

$$Na^{+-}[PhCHCH_2CH_2CHPh]^{-}Na^{+}$$

Note that the absence of both a termination and a transfer reaction means that if no accidental termination by impurity occurs the chains will remain active indefinitely.

The validity of this assumption has been demonstrated (i) by adding more styrene to the "living" polystyryl carbanions, and (ii) by adding another monomer such as isoprene, to form a block copolymer.

The existence of "living" polymers was originally demonstrated by Szwarc using an all-glass apparatus of the type shown in figure 4.1.

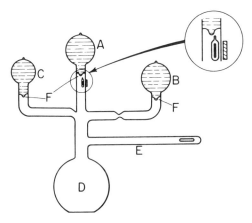

FIGURE 4.1. Apparatus similar to that used by Szwarc to demonstrate the existence of "living" polymers. The insert shows the arrangement of the internal and external magnets in relation to a break seal.

The components of the reaction were subjected to stringent purification procedures and sealed in the apparatus under vacuum. The green solution of the initiator, sodium naphthalene, in THF was contained in B and introduced into D by rupturing the break-seal using a glass-encased magnet with a sharp tip, contained in the apparatus. The magnet can be held in position by a second magnet taped in position on the exterior surface of the glass tubing. Pure styrene from C was then admitted to the reactor and an immediate colour change from green to red was observed which persisted after the rapid reaction was complete. The viscosity of the reaction mixture was tested by tipping it into the side arm E, after rotation to the vertical position, and timing the fall of a piece of metal encased in glass, through the medium. The apparatus was returned through 90° to its original position and a fresh solution of styrene in THF, having the same concentration as the reaction mixture, was added from bulb A. A marked increase in viscosity indicated further growth of the existing chains (rather than new ones being formed) and the red colour of the polystyryl ion was retained. In a second experiment, isoprene was contained in bulb A and when added, formed a block copolymer with the styrene. Analysis of the product showed that no polyisoprene was formed, again substantiating the concept of a "living" polymer. The use of "living" polymers is now a standard method of preparing block copolymers as well as the more unusual star-shaped and comb-branched polymers.

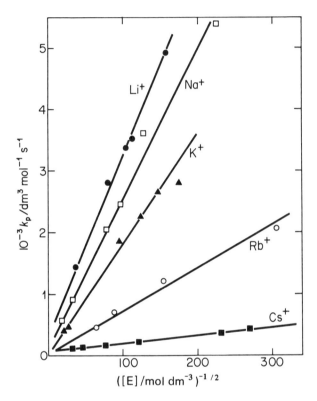

FIGURE 4.2. Behaviour of the experimental propagation rate constant k_p as a function of the concentration [E] of "living ends" for various salts of "living" polystyrene in tetrahydrofuran at 298 K. (From data by Szwarc.)

4.13 Metal alkyl initiators

The organo-lithium derivatives, such as n-butyl lithium, are particular members of this group of electron deficient initiators. In general, the initiation involves addition to the double bond of the monomer

$$\text{RLi} + \text{CH}_2\!\!=\!\!\text{CHR}_1 \rightarrow \text{RCH}_2\!-\!\!\overset{\displaystyle H}{\underset{\displaystyle R_1}{\overset{\displaystyle |}{\underset{\displaystyle |}{C}}}}{}^{\!-}\ \text{Li}^+$$

and propagation is then

$$\underset{\underset{R_1}{|}}{\overset{\overset{H}{|}}{RCH_2C^-}}\,Li^+ + nCH_2{=}CHR_1 \rightarrow R{-}(CH_2{-}CHR_1)_n\,\underset{\underset{R_1}{|}}{\overset{\overset{H}{|}}{CH_2C^-}}\,Li^+$$

Kinetic analysis of the reactions shows that the initiation is not a simple function of the basicity of R, however, owing to the characteristic tendency for organo-lithium compounds to associate and form tetramers or hexamers. The kinetics are usually complicated by this feature, which is solvent dependent, and consequently fractional reaction orders are commonplace.

The alkyl lithiums have proved commercially useful in diene polymerization and some steric control over the polymerization can be obtained.

4.14 Solvent and gegen-ion effects

Both the solvent and the gegen-ion have a pronounced influence on the rates of anionic polymerizations. The polymerization rate generally increases with increasing polarity of the solvent, for example, $k_p = 2.0$ dm^3 mol^{-1} s^{-1} for the anionic polymerization of styrene in benzene, but $k_p = 3800$ dm^3 mol^{-1} s^{-1} when the solvent is 1,2-dimethoxyethane. Unfortunately the dielectric constant

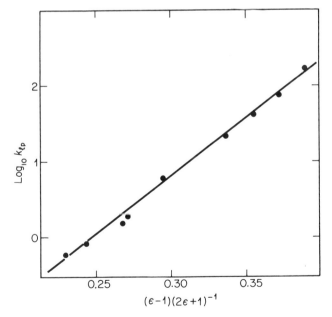

FIGURE 4.3. The propagation rate constant k_{lp} for the polystyryl-lithium ion pair plotted as a function of the dielectric constant of the reaction medium. (Adapted from Bywater and Worsfold.)

is not a useful guide to polarity or solvating power in these systems as k_p = 550 dm^3 mol^{-1} s^{-1} when the solvent is changed to THF whose dielectric constant ϵ is higher than ϵ for 1,2-dimethyoxyethane.

The influence of the gegen-ion on the polymerization of styrene in THF at 298 K is shown in figure 4.2 compiled from data obtained by Szwarc. Clearly the smaller Li$^+$ ions can be solvated to a greater extent than the larger ions and the decreasing rate reflects the increasing tendency for ion pairs to be the active species, rather than free ions, as the solvating power of the solvent deteriorates.

The effect of increasing the dielectric constant of the solvent on the propagation rate constant for an ion pair k_{lp}, has been demonstrated by Bywater and Worsfold (see figure 4.3). They plotted $\log_{10} k_{lp}$ against $(\epsilon - 1)/(2\epsilon + 1)$ for polystyryllithium in several THF + benzene mixtures and found that k_{lp} increased as the solvation increased. The solvation was measured by an increase in ϵ.

General Reading

A. M. Eastham, "Cationic polymerization" in *Encyclopaedia of Polymer Science and Technology*. Interscience Publishers Inc. (1965).

R. W. Lenz, *Organic Chemistry of Synthetic High Polymers*, Chapters 13 and 14. Interscience Publishers Inc. (1967).

D. Margerison and G. C. East, *Introduction to Polymer Chemistry*, Chapter 5. Pergamon Press (1967).

G. G. Overberger, J. E. Mulvaney and A. M. Schiller, "Anionic polymerization" in *Encyclopaedia of Polymer Science and Technology*. Interscience Publishers Inc. (1965).

P. H. Plesch, *The Chemistry of Cationic Polymerization*. Pergamon Press, 1963.

D. A. Smith, *Addition Polymers*, Chapter 3. Butterworths (1968).

M. Szwarc, *Carbanions, Living Polymers and Electron Transfer Processes*. Interscience Publishers Inc. (1968).

References

1. S. Bywater, "Polymerization initiated by lithium and its compounds", *Adv. in Polymer Science,* **4**, 66 (1965).
2. S. Bywater and D. J. Worsfold, *J. Phys. Chem.,* **70**, 162 (1966).
3. D. N. Bhattacharyya, C. L. Lee, J. Smid, and M. Szwarc, *J. Phys. Chem.,* **69**, 612, (1965).
4. J. P. Kennedy and A. W. Langer, "Recent advances in cationic polymerization", *Adv. in Polymer Science,* **3**, 508 (1964).

CHAPTER 5

Copolymerization

5.1 General characteristics

In the addition reactions considered in the previous chapters, the emphasis has been on the formation of a polymer from only one type of monomer. Often it is found that these homopolymers have widely differing properties and one might think that by using physical mixtures of various types, a combination of all the desirable properties would be obtained in the resulting material. Unfortunately this is not always so, and instead it is more likely that the poorer qualities of each become exaggerated in the mixture.

An alternative approach is to try to synthesize chains containing more than one monomer and examine the behaviour of the product. By choosing two (or perhaps more) suitable monomers, A and B, chains incorporating both can be prepared using free radical or ionic initiators, and many of the products exhibit the better qualities of the parent homopolymers. This is known as *copolymerization.*

Even in the simplest case, that of copolymerization involving two monomers, a variety of structures can be obtained, and five important types exist:

(i) *Random copolymers* are formed when irregular propagation occurs and the two units enter the chain in a random fashion, *i.e.* $\sim$ ABBAAAABAABBBA $\sim$. This is the most commonly encountered structure.

(ii) *Alternating copolymers* are obtained when equimolar quantities of two monomers are distributed in a regular alternating fashion in the chain $\sim$ ABABABA $\sim$. Many of the step-growth polymers formed by the condensation of two (A—A), (B—B) type monomers could be considered as alternating copolymers but these are commonly treated as homopolymers with the repeat unit corresponding to the dimeric residue.

(iii) *Block copolymers.* Instead of having a mixed distribution of the two units, the copolymer may contain long sequences of one monomer joined to another

sequence or block of the second. This produces a linear copolymer of the form
AA⏤AABBB⏤B, *i.e.* an {A} {B} block or sometimes an {A} {B} {A} type
block copolymer.

(iv) *Graft copolymers.* A non-linear or branched block copolymer is formed
by attaching chains of one monomer to the main chain of another homopolymer.

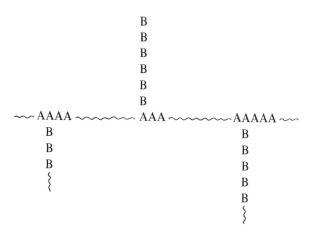

(v) *Stereoblock copolymers.* Finally a very special structure can be formed
from one monomer where now the distinguishing feature is the tacticity of
each block, *i.e.*

In general block and graft copolymers possess the properties of both homopolymers, whereas the random and alternating structures have characteristics
which are more of a compromise between the extremes.

It soon becomes obvious that the factors influencing the course of even simple
copolymerizations are much more complex than those in a homopolymerization.
For example, attempts to polymerize styrene and vinyl acetate result in copolymers containing only 1 to 2 per cent of vinyl acetate while a small quantity of
styrene will tend to inhibit the free radical polymerization of vinyl acetate. At
the other extreme, two monomers like maleic anhydride and stilbene are
extremely difficult to polymerize separately, but form copolymers with relative
ease.

5.2 Composition drift

It was realized by Staudinger, as early as 1930, that when two monomers
copolymerize the tendency of each monomer to enter the chain can differ
markedly. He found that if an equimolar mixture of vinyl acetate and vinyl

chloride were copolymerized, the chemical composition of the product varied throughout the reaction, and that the ratio of chloride to acetate in the copolymers changed from 9:3 to 7:3 to 5:3 to 5:7.

This phenomenon, known as *composition drift*, is a feature of many copolymerizations and has been attributed to the greater reactivity of one of the monomers in the mixture. Consequently, in a copolymerization, it is necessary to distinguish between the composition of a copolymer being formed at any one time in the reaction and the overall composition of all the polymer formed at a given degree of conversion.

Two major questions arise which must be answered if the criteria controlling copolymerizations are to be formulated.

(1) Can the composition of the copolymer be predicted when it is prepared from the restricted conversion of a mixture of two monomers?

(2) Can one predict the behaviour of two monomers which have never reacted before?

To answer the first question, we must explore the relative reactivity of one monomer to another, while an attempt to answer the second is embodied in the $Q-e$ scheme.

5.3 The copolymer equation

To begin to answer question (1) we must establish a suitable kinetic scheme. The following group of homo- and hetero-polymerization reactions were proposed by Dostal in 1936 for a radical copolymerization between two monomers M_1 and M_2, and ultimately extended and formalized by a number of workers who established a practical equation from the reactions:

$$\sim M_1^{\cdot} + M_1 \xrightarrow{k_{11}} \sim M_1^{\cdot}$$

$$\sim M_1^{\cdot} + M_2 \xrightarrow{k_{12}} \sim M_2^{\cdot}$$

$$\sim M_2^{\cdot} + M_2 \xrightarrow{k_{22}} \sim M_2^{\cdot}$$

$$\sim M_2^{\cdot} + M_1 \xrightarrow{k_{21}} \sim M_1^{\cdot}$$

where k_{11} and k_{22} are the rate constants for the *self-propagating* reactions and k_{12} and k_{21} are the corresponding *cross-propagation* rate constants.

Under steady-state conditions, and assuming that the radical reactivity is independent of chain length and depends only on the nature of the terminal unit, the rate of consumption of M_1 from the initial reaction mixture is then

$$- d[M_1]/dt = k_{11}[M_1][M_1^{\cdot}] + k_{21}[M_1][M_2^{\cdot}], \qquad (5.1)$$

and M_2 by

$$- d[M_2]/dt = k_{22}[M_2][M_2^{\cdot}] + k_{12}[M_2][M_1^{\cdot}]. \qquad (5.2)$$

The *copolymer equation* can then be obtained by dividing equation (5.1) by (5.2) and assuming that $k_{21}[M_2^*][M_1] = k_{12}[M_1^*][M_2]$ for steady-state conditions, so that

$$d[M_1]/d[M_2] = ([M_1]/[M_2])\{(r_1[M_1] + [M_2])/([M_1] + r_2[M_2])\}, \qquad (5.3)$$

where $k_{11}/k_{12} = r_1$, and $k_{22}/k_{21} = r_2$.

The quantities r_1 and r_2 are the relative reactivity ratios defined more generally as the ratio of the reactivity of the propagating species with its own monomer to the reactivity of the propagating species with the other monomer.

5.4 Monomer reactivity ratios

The copolymer equation provides a means of calculating the amount of each monomer incorporated in the chain from a given reaction mixture or feed, when the reactivity ratios are known. It shows that if monomer M_1 is more reactive than M_2, then M_1 will enter the copolymer more rapidly, consequently the feed becomes progressively poorer in M_1 and composition drift occurs. The equation is then an "instantaneous" expression which relates only to the feed composition at any given time.

TABLE 5.1. Some reactivity ratios r_1 and r_2 for free radical initiated copolymerizations

M_1	M_2	r_1	r_2	$r_1 r_2$
Acrylonitrile	Acrylamide	0.87	1.37	1.17
	Butadiene	2.0	0.1	0.2
	Methyl acrylate	0.84	0.83	0.70
	Styrene	0.01	0.40	0.004
	Vinyl acetate	6.0	0.07	0.42
Butadiene	Methyl methacrylate	0.70	0.32	0.22
	Styrene	1.40	0.78	1.1
Ethylene	Propylene	17.8	0.065	1.17
Maleic anhydride	Acrylonitrile	0	6	0
	Methyl acrylate	0.02	3.5	0.07
Methyl methacrylate	Vinyl acetate	22.2	0.07	1.55
	Vinyl chloride	10	0.1	1.0
Styrene	*p*-Fluorostyrene	1.5	0.7	1.05
	α-Methylstyrene	2.3	0.38	0.87
	Vinyl acetate	55	0.01	0.55
	2-Vinyl pyridine	0.55	1.14	0.63
Tetrafluoroethylene	Monochlorotrifluoroethylene	1.0	1.0	1.0
Vinyl chloride	Vinyl acetate	1.35	0.65	0.88
	Vinylidene chloride	0.5	0.001	0.0005

As r_1 and r_2 are obviously the factors which control the composition of the copolymer, one must obtain reliable values of r for each pair of monomers (comonomers) if the copolymerization is to be completely understood and controlled. This can be achieved by analysing the composition of the copolymer formed from a number of comonomer mixtures with various $[M_1]/[M_2]$ ratios, at low (5 to 10 per cent) conversions (where monomer reactivities do not differ greatly).

If we now define F_1 and F_2 as the mole fractions of monomers M_1 and M_2 being added to the growing chain at any given time, and f_1 and f_2 the corresponding mole fractions of the monomers in the feed, then $(d[M_1]/d[M_2]) = (F_1/F_2) = F$ and $([M_1]/[M_2]) = (f_1/f_2) = f$. Substitution into equation (5.3) leads to a simplified form of the copolymer equation

$$\{f(1 - F)/F\} = r_2 + (f^2/F)r_1, \tag{5.4}$$

and a plot of $\{f(1 - F)/F\}$ against (f^2/F) will give a straight line having r_1 as the slope and r_2 as the intercept.

Some representative values of r_1 and r_2 are shown in table 5.1 for a number of comonomers, and they are seen to differ widely.

5.5 Reactivity ratios and copolymer structure
It is obvious, from the wide ranging values of reactivity ratios shown in table 5.1, that the structure of the copolymer will also be a function of r_1 and r_2.

Several types of copolymeric structure can be obtained as we saw in section 5.1 and the influence of monomer reactivity ratios can be illustrated by examining plots of the "instantaneous" copolymer composition F_1 against the "instantaneous" monomer composition in the feed f_1, for various combinations of r_1 and r_2.

Consider first the unusual case when $r_1 \approx r_2 \approx 1$. This situation arises when little or no preference for either monomer is shown by the polymer radical, i.e. $k_{11} \approx k_{12}$ and $k_{22} \approx k_{21}$, and copolymerization is then entirely random. Under these conditions $F_1 = f_1$ and this is represented by curve I in figure 5.1. As this plot is reminiscent of corresponding plots for an ideal system of two liquids, the copolymers formed under these conditions, and indeed any copolymer where the product (r_1r_2) is unity, are called IDEAL copolymers. Completely random copolymers are formed from the comonomer pairs: tetrafluoroethylene + monochlorotrifluoroethylene; isoprene + butadiene; vinyl acetate + isopropenyl acetate. However, if $r_1 > 1$ and $r_2 < 1$ or vice versa, but $r_1r_2 = 1$, there will be composition drift of the kind shown in figure 5.2, and when the differences between r_1 and r_2 become large, departure from ideal conditions is significant. The curve for $r_1 = 5.0$ and $r_2 = 0.2$ clearly shows that M_1 enters the copolymer more frequently than M_2 and random copolymers become increasingly difficult to prepare.

Values of (r_1r_2) are, however, more likely to be above or below unity and curve II, of figure 5.1 represents the nearly ideal pair acrylamide + acrylonitrile, for which $r_1 r_2 = 1.17$. This shows the slight deviation from ideal copolymerization and illustrates the use of the curve as a guide to the composition drift which can be expected when $r_1 \neq r_2$.

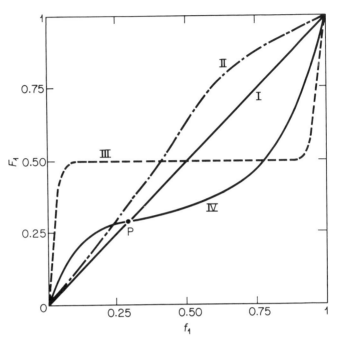

FIGURE 5.1. Variation of F_1 with f_1 for copolymerizations which are: I, completely random; II, almost ideal (i.e. $r_1 r_2 = 1.17$); III, regular alternating; IV, intermediate between alternating and random (*i.e.* $0 < r_1 r_2 < 1$).

In systems where r_1 and r_2 are both less than unity, copolymerization is favoured and only short sequences of M_1 and M_2 tend to form. In the extreme case when k_{11} and k_{22} are zero, $r_1 = r_2 = 0$ and a regular alternating (1:1) copolymer is formed; this is represented by curve III of figure 5.1. Strictly alternating copolymers can be prepared from the comonomers maleic anhydride + styrene, fumaronitrile + α-methyl styrene and others; however, these are rather special cases and more generally there is a greater likelihood for systems lying in the range $0 < r_1 r_2 < 1$. Thus the closer the product $(r_1 r_2)$ is to zero, the greater is the tendency for M_1 and M_2 to alternate in the chain. The copolymer composition plots for these types of system are sigmoidal (curve IV) and cross the ideal line at a point P. At this point $F_1 = f_1$, and P indicates the *azeotropic copolymer composition.* This is an important feature of the system, as it represents a feed composition which will produce a copolymer of constant composition throughout the whole reaction, without having to make adjustment to the feed. This type of copolymerization, where

no composition drift is observed, is known as *azeotropic* copolymerization
and the critical composition f_{1c} required to obtain the necessary conditions
can be calculated from

$$f_{1c} = (1 - r_2)/\{2 - (r_1 + r_2)\}. \tag{5.5}$$

When r_1 and r_2 are greater than unity, *i.e.* $r_1 r_2 \gg 1$, conditions favouring
long sequences or blocks of each monomer in the copolymer are obtained, and,
in extreme cases, homopolymer formation may predominate.

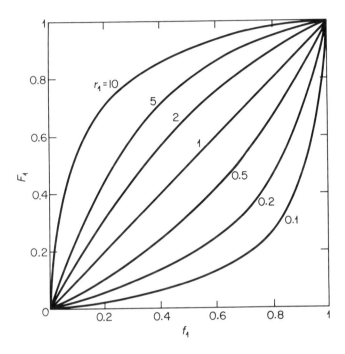

FIGURE 5.2. Plot of mole fraction F_1 of comonomer 1 in the copolymer as a
function of the mole fraction f_1 of comonomer 1 in the feed, for copolymerizations in which $r_1 r_2 = 1$, showing compositional variations for several indicated
values of r_1.

5.6 Monomer reactivities and chain initiation

Monomer reactivities have been found to be essentially independent of the free
radical process used (*e.g.* bulk, emulsion) but can be affected tremendously, for
the same pair of monomers, if the chain carrier is changed.

For example, monomer reactivity ratios for styrene and methyl methacrylate
in a free radical copolymerization are $r_1 = 0.5, r_2 = 0.44$. This represents a
random copolymerization. Contrast this with the anionic reaction, where $r_1 = 0.12$
and $r_2 = 6.4$, or the cationic reaction where $r_1 = 10.5$ and $r_2 = 0.1$. Obviously

the propagation rates are no longer similar and this is represented in figure 5.3 where it can be seen that the anionic technique produces a copolymer rich in methyl methacrylate while the cationic system leads to a copolymer with a high styrene content.

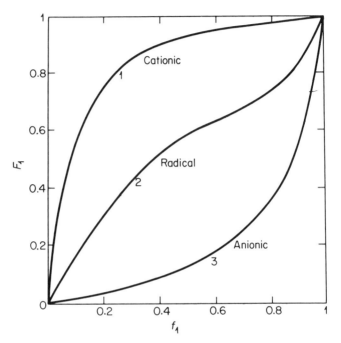

FIGURE 5.3. Copolymerization of styrene and methyl methacrylate initiated by 1, $SnCl_4$; 2, benzoyl peroxide; and 3, sodium in liquid ammonia, showing the vast differences in the dependence of F_1 on f_1 for the various types of initiator, where component 1 is styrene.(After Pepper.)

This illustration merely accentuates the need to answer the questions, why do the values of r_1 and r_2 differ so widely and why does r for a given monomer change when the comonomer is changed?

5.7 Influence of structural effects on monomer reactivity ratios

The propagation rates in ionic polymerizations are influenced by the polarity of the monomers; in free radical reactions the relative reactivity of the monomers can be correlated with resonance stability, polarity, and steric effects; we shall consider only radical copolymerizations.

Resonance effects. The reactivity of a free radical is known to depend on the nature of the groups in the vicinity of the radical. If, in a vinyl monomer $(CH_2\!=\!CHR)$, the group R is capable of aiding delocalization of the radical, the radical stability will increase, and some of the more common substituents can be arranged in order of increasing electron withdrawal:

$$C_6H_5 > -CH\!\!=\!\!CH_2 > -\overset{\displaystyle |}{\underset{\displaystyle O}{\overset{\displaystyle \|}{C}}}-CH_3 > C\!\!\equiv\!\!N > -\overset{\displaystyle \|}{\underset{\displaystyle O}{C}}-OR > Cl > R$$

$$> -O-\overset{\displaystyle \|}{\underset{\displaystyle O}{C}}-CH_3$$

Thus styrene (R = C_6H_5) has a radical whose resonance stabilization is high (84 kJ mol^{-1}) whereas vinyl acetate (R = O$-\overset{\displaystyle \|}{\underset{\displaystyle O}{C}}-CH_3$) has a very unstable radical.

As a reactive monomer forms a stable free radical, the radical reactivity will be the reverse order of the groups above. This means that monomers containing conjugated systems (styrene, butadiene, acrylates, acrylonitriles, *etc.*) will be highly reactive monomers but will form stable and so relatively unreactive radicals. Conversely, unconjugated monomers (ethylene, vinyl halides, vinyl acetate, *etc.*) are relatively unreactive towards free radicals but will form unstable and highly reactive adducts.

The suppression of radical reactivity towards a monomer is also found to be a stronger effect than the corresponding enhancement of monomer reactivity. This is true for styrene whose radical is about 10^3 times less reactive towards a given monomer than the vinyl acetate radical, but the styrene monomer is only 50 times more reactive towards a given radical than the vinyl acetate monomer.

We can now see why styrene and vinyl acetate are such a poor comonomer pair. The copolymerization requires that the stable styrene radical reacts with the unreactive vinyl acetate monomer, but this is such a slow process that the styrene tends to homopolymerize.

Broadly speaking, an efficient copolymerization tends to take place when the comonomers are either both reactive or both relatively unreactive, but not when one is reactive and the other unreactive. As with most generalizations, this is rather an extreme statement and cannot be treated too rigorously, especially when one realizes that resonance is not the only factor contributing to copolymerization behaviour, and that both steric and polar effects have to be considered.

Polar effects. It has been observed that strongly alternating copolymers are formed when comonomers with widely differing polarities are reacted together. The polarity is again determined by the side group. Thus electron withdrawing substituents, *e.g.* —COOR, —CN, —COCH$_3$, all decrease the electron density of the double bond in a vinyl monomer relative to ethylene, whereas electron donating groups, *e.g.* —CH$_3$, —OR, —OCOCH$_3$ increase the electron density.

Hence acrylonitrile forms random copolymers with methyl vinyl ketone ($r_1 r_2 = 1.1$), while copolymerization of acrylonitrile with vinyl ethers leads to alternating structures ($r_1 r_2 \approx 0.0004$).

Polar forces also help to overcome steric hindrance. Neither maleic anhydride nor diethyl fumarate will form homopolymers, but both will react with styrene, stilbene, and vinyl ethers to form alternating copolymers because of the strong polar interaction. For example, the reaction between stilbene and maleic anhydride is

The Q - e scheme

All these factors contribute to the rate of copolymerization, but in a manner which makes it difficult to distinguish the magnitude of each effect.

Attempts to correlate copolymerization tendencies are thus mainly on a semi-empirical footing and must be treated as useful approximations rather than rigorous relations. A generally useful scheme was proposed by Alfrey and Price who denoted the reactivities or resonance effects of monomers by a quantity Q and radicals by P, while the polar properties were assigned a factor e which is assumed to be the same for both a monomer and its radical.

An expression for the rate constant of the cross-propagation reaction can then be derived as

$$k_{12} = P_1 Q_2 \exp(-e_1 e_2), \tag{5.6}$$

where P_1 relates to the radical M_1^* and Q_2 to the monomer M_2. This has been called the $Q - e$ scheme and can be used to predict monomer reactivity ratios by extending the treatment to give the relations for r_1 and r_2:

$$r_1 = (k_{11}/k_{12}) = (Q_1/Q_2) \exp\{-e_1(e_1 - e_2)\}, \tag{5.7}$$

$$r_2 = (k_{22}/k_{21}) = (Q_2/Q_1) \exp\{-e_2(e_2 - e_1)\}, \tag{5.8}$$

and

$$r_1 r_2 = \exp\{-(e_1 - e_2)^2\}. \tag{5.9}$$

By choosing arbitrary reference values for styrene of $Q = 1.0$ and $e = -0.8$, a
table of relative values of Q and e for monomers can be compiled.

TABLE 5.2. Selected values of Q and e for monomers

Monomer	Q	e
Styrene (reference)	1.0	-0.8
Acrylonitrile	0.60	1.20
1,3-Butadiene	2.39	-1.05
Isobutylene	0.033	-0.96
Ethylene	0.015	-0.20
Isoprene	3.33	-1.22
Maleic anhydride	0.23	2.25
Methyl methacrylate	0.74	0.40
α-Methyl styrene	0.98	-1.27
Propylene	0.002	-0.78
Vinyl acetate	0.026	-0.25
Vinyl chloride	0.044	0.20

On doing this one finds that for substituents capable of conjugating with the
double bond $Q > 0.5$, whereas for groups such as Cl, OR, and alkyl, $Q < 0.1$,
thereby reflecting the assumption that Q is a measure of resonance stabilization.

The values of e are also informative; for instance, maleic anhydride with two
strong electron attracting side groups has $e = +1.5$ indicating an electropositive
double bond. This leads to a repulsion of other maleic anhydride molecules and
so no homopolymerization takes place. Similarly isobutylene has $e = -1.1$, and
repulsion of like monomers is again a strong possibility. Copolymerization of
oppositely charged monomers, however, should take place readily.

Although the scheme suffers from the disadvantages that steric effects are
ignored, that the use of the same value of e for both monomer and radical is a
doubtful assumption, and that monomers other than monosubstituted ethylenes
do not fit in very well, it has proved useful in a qualitative way and should be
accepted for what it is – a useful approximation.

The equation is similar to the Hammett equation which correlates monomer
reactivity with structure, but the Hammett treatment is limited to substituted
aromatic compounds.

5.9 Block copolymers

Copolymers prepared by linking two or more long linear sequences of different
homopolymers together are called block copolymers. If $\{A_n\}$ represents a sequence
n units long, then the following types may be prepared: a di-block $\{A_n\}\{B_m\}$;
a tri-block or sandwich block $\{A_n\}\{B_m\}\{A_p\}$; a tri-block $\{A_n\}\{B_m\}\{C_p\}$,
or if the blocks are composed of the same monomer in different steric configurations, a stereoblock.

"Living" polymers are most useful in the synthesis of block copolymers and the technique offers the additional advantage that the block size can be controlled when conditions are right. A simple di-block can be prepared when the "living" polymer of monomer A is able to initiate the polymerization of an added monomer B, to give an $\{A_n\}$ $\{B_m\}$ block. Initiating abilities vary, however, and while the "living" polystyryl anion initiates the polymerization of methyl methacrylate to give poly(styrene-b-methyl methacrylate), the reverse is not true. Similarly the polystyryl anion will initiate the polymerization of ethylene oxide but again *not vice versa.* These reactivities have been related to the values of Q and e and it is suggested that the anion of a monomer with a low value of e will initiate polymerization of a monomer with a higher e, but that the reverse is not true.

"Living" polymers which propagate from both ends, the di-anions, are useful in the preparation of $\{A\}$ $\{B\}$ $\{A\}$ blocks.

Polymers with other types of active end groups can also be used, such as hydroxyl, amine, and carboxyl terminated polymers. These are easily utilized in step-growth reactions but can be extended to other systems. Poly(ethylene oxides) will react with propylene oxide to form surface active block copolymers called "Pluronics". Vinyl polymers with hydroxyl end groups can be prepared if initiated by hydrogen peroxide and these can be reacted subsequently with

$$H_2O_2 + Fe^{2+} \rightarrow Fe^{3+} + OH^- + HO^{\cdot} + CH_2 {=\!\!=} CHX \rightarrow HO \left(CH_2CHX \right)_n$$

other blocks. Elastomeric polymers are synthesized when diisocyanates react with hydroxyl terminated polyesters. A segmented polyurethane polymer results from the reaction between diisocyanates and hydroxyl terminated polyethers, and consists of alternate "hard" and "soft" sequences joined together in a linear chain. The polyurethane blocks provide the "hard" or crystalline segments, with high melting temperatures. The low melting temperature "soft" segments, whose glass transition temperatures are well below ambient temperatures, are composed of polyether blocks, typically poly(tetramethylene oxide). These provide the rubber-like portions of the polymer. The copolymers can be spun into elastomeric fibres, such as "Spandex", and find uses in swimwear, surgical hosiery, *etc.,* where improved stretch and shape recovery fabrics are required.

If a "living" polymer is not readily available, the required polymer can often be prepared with an end group suitable for radical reactions. The solvent chain transfer reaction which occurs when styrene is polymerized in CBr_4 forms polymers with terminal bromine. The photochemical removal of the halogen by irradiation with u.v. light in the presence of methyl methacrylate initiates free radical polymerization at the chain end to form an $\{A\}$ $\{B\}$ block. Alternatively styrene can be polymerized under mild initiation conditions with

phthaloyl peroxide

when a two-stage decomposition

of the initiator takes place. A chain incorporating the undecomposed peroxide group after the first stage can be reactivated under more rigorous conditions in the presence of a second monomer to produce an $\{A\}$ $\{B\}$ block. Many other examples exist.

Block copolymers tend to behave more like blends of the homopolymers when the blocks are long and they may exhibit properties reminiscent of the parent chains. The blocks often display a certain degree of incompatibility which leads to interesting and often useful behaviour. For example $\{A\}$ $\{B\}$ blocks of polystyrene and polyethylene oxide form gels when mixed with either butyl phthalate, which is a good solvent for polystyrene but a precipitant for polyethylene oxide, or nitromethane, where the reverse is true. These show lamellar or micellar structures respectively, caused by the aggregation of the $\{A\}$ blocks with other $\{A\}$ blocks in the various chains and the $\{B\}$ with $\{B\}$. These copolymers with polar and non-polar block combinations would be expected to make efficient homogenizing agents and can be used in polymer solution + oil mixtures.

This ordering process in block copolymers can produce spectacular visual effects and concentrated solutions of high molar mass styrene + butadiene blocks show highly irridescent colour changes on the application of a stress. The colours also change with the amount of solvent because the display results from light interference in the layers formed by the incompatible blocks and the spacings are dependent upon the concentration.

Sandwich blocks of styrene with either isoprene or butadiene have produced an industrially interesting group of thermoplastic elastomers, with the polystyrene blocks acting as both filler and crosslinking agent at temperatures below its glass transition. These are discussed in chapter 14.

5.10 Graft copolymers

Whereas block copolymers are linear chains formed by introducing active sites in the terminal units, graft copolymers are their branched equivalents where the active site is included on an internal monomer unit. Grafting can be effected by a free radical initiated mechanism. This usually involves hydrogen abstraction from the main chain in the presence of another monomer which then grows from this site as a branch. The reaction is known as *transfer* grafting. Unfortunately, the products often contain homopolymer and unmodified chains in addition to the graft product, and the efficiency of the process depends on a number of factors, the most important of which is the initiator. This is demonstrated in the system poly(styrene–g-methyl methacrylate). Benzoyl peroxide (BPO) is a good initiator, but neither azobisisobutyronitrile (AIBN) nor t-butyl peroxide show any tendency to initiate the graft. On the other hand, the grafting efficiency for poly(methyl methacrylate–g-vinyl acetate) is apparently independent of the initiator, while styrene or methyl methacrylate will graft on to polyisoprene using BPO but not AIBN. This behaviour may be a result of resonance stabilization of the nitrile radical.

Grafting tendencies are similar to copolymerization behaviour and values of r are useful in this respect; thus the grafting of vinyl acetate on polystyrene $r_1 = 0.01, r_2 = 55$ is poor compared with its effective reaction with poly(vinyl chloride) $r_1 = 0.4, r_2 = 1.7$.

Activation grafting takes place when there is absorption of radiation to create an active site on the chain and two techniques are employed. (i) Pre-irradiation involves exposure of a polymer to the source of energy before addition of the monomer, but as this requires radicals which are either trapped or have long lives it is not commonly used. (ii) Mutual irradiation, where monomer and polymer are mixed prior to the reaction is preferable. If the energy source is ionizing radiation, non-selective bond cleavage is obtained and the effectiveness may depend on the sensitivity of the monomer to the radiation. This can be estimated from the value of G, which is a measure of the number of radicals formed per 100 eV absorbed. Good grafting combinations result from monomers with low values of G and polymers with high values of G. Greater control over the graft process can be achieved if photolysis of a carbonyl group or a halogen in a side group is feasible.

Chemical grafting is also possible and the redox reaction of pendant hydroxyls with cerium(IV) ions is commonly used.

$$\sim\!\!\sim CH_2\!\!-\!\!CH\!\!\sim\!\!\sim + Ce^{4+} \rightarrow \sim\!\!\sim CH_2CH\!\!\sim\!\!\sim + H^+ + Ce^{3+}$$
$$\underset{\text{ROH}}{|} \qquad\qquad\qquad \underset{\text{RO}^{\bullet}}{|}$$

If carried out in the presence of a monomer, grafting is readily accomplished and this is particularly useful in the preparation of cellulose graft copolymers. Anionic grafting has also been effected.

Copolymers are prepared to gain a wider variety of properties and in certain situations may be much better than a homopolymer. Many synthetic fibres are resistant to dyeing and the response of polyamides can be improved by grafting some acrylic acid chains on to the polymer at the site of the active hydrogen. Poly(acrylonitrile-*co*-vinyl pyridine) fibres are also more amenable to dyeing than pure polyacrylonitrile, while improved heat resistance is obtained with materials made from poly(methyl methacrylate-*co*-styrene) rather than poly-(methyl methacrylate) alone. The thermal stability of poly(vinyl chloride) is also vastly improved by copolymerization with propylene, and a polymer which is tougher and more flexible than polyethylene is obtained in poly(ethylene-*co*-vinyl acetate). Many other examples exist but a group known as "ionomers" have proved particularly interesting.

5.11 Ionomers

Ionomers are prepared by the random copolymerization of an α-olefin $CH_2\!=\!CR_1R_2$, where R_1 is H, alkyl, or COOH and R_2 is COOH, with a carboxylic acid. The ionomer is formed when the free carboxyl groups are partially neutralized by a reaction with a metal cation. This increases the associative forces between carboxyl groups in juxtaposition and creates ionic crosslinks between the chains when they form an entangled network.

The polymer structure is comprised of three phases and can be represented by the schematic diagram in figure 5.4. X-ray analysis suggests that the cations exist as spherical clusters, about 10 nm diameter, in conjunction with ordered crystalline regions, both of which are embedded in an amorphous matrix. The crystallites are represented as regularly folded sections with the disordered amorphous regions acting as the third interconnecting phase. Individual cations are thought to be spaced about 2 nm apart.

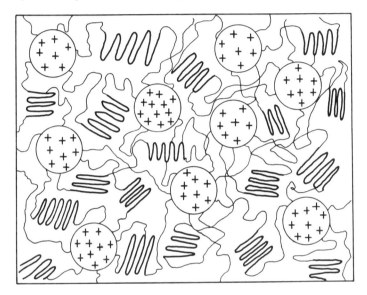

FIGURE 5.4. Schematic representation of the three-phase structure of a dry ionomer, consisting of cation clusters, lamellae, and disordered regions.

Prior to the addition of the cations, the polymers are highly crystalline and consequently opaque, but the long range order is partially destroyed by the ionic crosslinking and the polymer becomes transparent.

The crosslinking imparts typical elastomeric properties to the ionomers but because they are weak electrostatic rather than covalent links, the polymers are also thermoplastic. They are sometimes called "elastoplastic".

Ionomers such as poly(ethylene-co-methacrylic acid), when treated with sodium or magnesium salts, act like thermosetting plastics at low temperatures and are used for films, tubing, and injection moulded articles. Property variations can be introduced by altering the molar mass and also the metal cation. An increase in the ionization leads to a clearer, stiffer, and tougher product, whereas a change of cation from Zn^{2+} to Li^+ to Na^+ produces an increasing resistance to stress cracking. This effect can be duplicated by using only one type of cation and increasing the molar mass instead.

A range of polyurethane ionomers can be formed by reacting diisocyanate prepolymer, prepared from polyester or polyether diols and excess diisocyanate, with salts of aliphatic diamino acids. The polyaddition is rapid and high tensile strength, solvent resistant ionomers are produced.

General Reading

S. L. Aggarwal, *Block Polymers.* Plenum Press (1970).

R. J. Ceresa, *Block and Graft Copolymers.* Butterworths (1962).

G. Odian, *Principles of Polymerization,* Chpater 6. McGraw-Hill (1970).

References

1. G. M. Estes, S. L. Cooper and A. B. Tobolsky, "Block copolymers", *Reviews in Macromolecular Chemistry,* **5**–2, 167 (1970).
2. D. C. Pepper, *Quarterly Reviews,* **8**, 88 (1954).

CHAPTER 6

Polymer Stereochemistry

The physical behaviour of a polymer depends not only on the general chemical composition but also on the more subtle differences in microstructure which are known to exist. As it is now possible to exercise a large degree of control over the synthesis of specific structures it is prudent at this point to elaborate on the types of microstructural variations encountered before discussing how each can be produced. Several kinds of isomerism or microstructural variations can be identified and these are grouped under four main headings: architectural, orientational, configurational, and geometric.

6.1 Architecture

Differences here include branching, network formation, and polymers derived from isomeric monomers, for example, poly(ethylene oxide), I, poly(vinyl alcohol), II, and polyacetaldehyde, III where the chemical composition of the monomer units is the same but the atomic arrangement is different in each case. This makes a considerable difference to the physical properties of the polymers, *e.g.* the glass transition temperature T_g of structure I is 206 K, for II T_g = 358 K, and for III T_g = 243 K.

$$
\overset{\text{I}}{-(\text{CH}_2\text{CH}_2-\text{O})_n} \qquad \overset{\text{II}}{\left(-\text{CH}_2-\underset{\overset{|}{\text{OH}}}{\text{CH}}-\right)_n} \qquad \overset{\text{III}}{\left(-\underset{\overset{|}{\text{CH}_3}}{\text{CH}}-\text{O}-\right)_n}
$$

6.2 Orientation

When a radical attacks an asymmetric vinyl monomer two modes of addition are possible

$$R^{\bullet} + CH_2\!=\!CH \quad\longrightarrow\quad
\begin{cases}
RCH_2CH^{\bullet} \quad\ \text{I} \\
\quad\ \ |\\
\quad\ \ X \\
\\
RCHCH_2^{\bullet} \quad \text{II} \\
\quad\ |\\
\quad\ X
\end{cases}
$$

with the monomer bearing substituent X.

This leads to the configuration of the monomer unit in the chain being either head-to-tail

$$\sim CH_2-\underset{\underset{X}{|}}{CH}-CH_2-\underset{\underset{X}{|}}{CH}-CH_2-\underset{\underset{X}{|}}{CH}-CH_2-\underset{\underset{X}{|}}{CH}\sim$$

III

if route I is favoured, or a chain containing a proportion of head-to-head, tail-to-tail structure IV if route II is followed.

$$\sim CH_2-\underset{\underset{X}{|}}{CH}-\underset{\underset{X}{|}}{CH}-CH_2-CH_2-\underset{\underset{X}{|}}{CH}-\underset{\underset{X}{|}}{CH}-CH_2-CH_2-\underset{\underset{X}{|}}{CH}\sim$$

head-head   IV   tail-tail

The actual mode of addition depends on two factors: the stability of the product and the possible steric hindrance to the approach of $R^{\bullet}$ caused by a large group X in the molecule. The reaction in route I is highly favoured, because firstly there is a greater possibility of resonance stabilization of this structure by interaction between group X and the unpaired electron on the adjacent α-carbon atom, and secondly this direction of radical attack is least impeded by the substituent X. The preferred structure is then the head-to-tail orientation (III) and while the alternative structure IV may occur occasionally in the chain, especially when termination by combination predominates, the existence of an exclusively head-to-head orientation is unlikely unless synthesized by a special route.

Experimental evidence supports the predominance of structure III in the majority of polymers; the most notable exceptions are poly(vinylidene fluoride) with 4 to 6 per cent and poly(vinyl fluoride) with 25 to 32 per cent head-to-head links detected by n.m.r. studies. The presence of head-to-tail structures can be demonstrated in a number of ways and the general principle is illustrated using poly(vinyl chloride) as an example. Treatment of this polymer with zinc dust in dioxan solution leads to elimination of chlorine which can proceed by two mechanisms.

(a) $\sim$ CH$_2$—CH—CH$_2$—CH—CH$_2$ $\sim$ $\rightarrow$ $\sim$ CH$_2$—CH—CH—CH$_2$ $\sim$
$\qquad\quad\;$ | $\qquad\quad$ | $\qquad\qquad\qquad\qquad$ \\ /
$\qquad\quad$ Cl $\qquad\quad$ Cl $\qquad\qquad\qquad\qquad\;\;$ CH$_2$
$\qquad\qquad\qquad\qquad\qquad\qquad\qquad\qquad$ + ZnCl$_2$

(b) $\sim$ CH$_2$—CH—CH—CH$_2$ $\sim$ $\rightarrow$ $\sim$ CH$_2$—CH=CH—CH$_2$ $\sim$ + ZnCl$_2$
$\qquad\quad\;$ | $\quad$ |
$\qquad\quad$ Cl $\;\;$ Cl

Statistical analysis of chlorine loss via route (a) indicates that only 86.4 per cent of the chlorine will react due to the fact that, as elimination is a random process, about 13.6 per cent of the chlorine atoms become isolated during the reaction and will remain in the chain. Elimination by mechanism (b) results in total removal of chlorine. Analysis of poly(vinyl chloride) after treatment with zinc dust showed 84 to 86 per cent chlorine elimination and this figure remained constant even after prolonged heating of the reaction mixture. This leads one to the conclusion that the polymer is almost entirely in the head-to-tail orientation.

6.3 Configuration

It has long since been recognized that when an asymmetric vinyl monomer CH$_2$=CHX is polymerized, every tertiary carbon atom in the chain can be regarded as a chiral centre by virtue of the fact that m and n are not normally equal in any chain. Under these circumstances the two possible configurations shown (i) and (ii) can only be interconverted by breaking a bond.

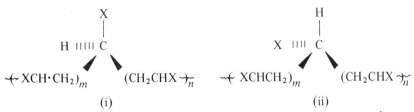

$\qquad\qquad\qquad\quad$ (i) $\qquad\qquad\qquad\qquad\qquad\qquad$ (ii)

No real progress in the preparation of distinguishable stereoisomers was made until the advent of the Ziegler-Natta catalysts which will be discussed later Since then the study of stereoregular polymers has expanded rapidly in a vigorous and exciting manner, helped greatly by the application of n.m.r. to the accurate characterization of the microstructure, but before examining these topics, a brief outline of the nomenclature is required.

If every tertiary carbon atom in the chain is asymmetric one might expect the polymer to exhibit optical activity. Normally homoatomic carbon chains show no optical activity because two long chains constitute part of the group variations and as these become longer (and more alike) in relation to the chiral centre, the optical activity decreases to a vanishingly small value. Vinyl polymers derived from (CH_2=CXY) monomers fall into this category as they are centro-symmetric relative to the main chain, and the tertiary carbons are then only pseudo-asymmetric.

This is not true for heteroatomic chains such as ~(CH_2C*HX·O)~ where C* is a true asymmetric centre, and these polymers are optically active. In this case an absolute configuration can be assigned, using preferably the Cahn-Ingold-Prelog system, referring either to the R- (rectus) or to the S- (sinister) form.

The two forms (i) and (ii) can be distinguished by arbitrarily assigning them d- or l-configurations, which have nothing to do with optical activity and merely refer to the group X being positioned either below or above the chain in a planar projection.

There are then three distinctive distributions of the d- and l-forms among the units in a chain and these decide the chain tacticity.

MONOTACTIC POLYMERS

(a) The *Isotactic* form. When a polymer, in the all *trans* zig-zag conformation, is viewed along the bonds comprising the chain backbone, then if each asymmetric chain atom has its substituents in the same steric order the polymer is said to be *isotactic*. In other words the arrangements of the substituent groups is either all *d* or all *l*.

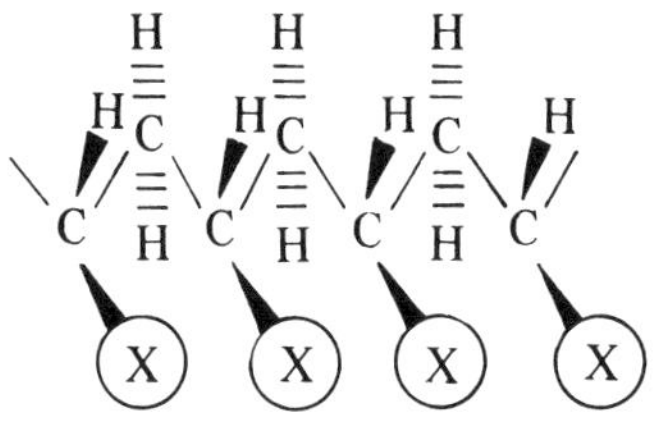

dddd

(b) *Syndiotactic.* A chain is termed syndiotactic when observation along the main chain shows the opposite configuration around each successive asymmetric centre in the chain.

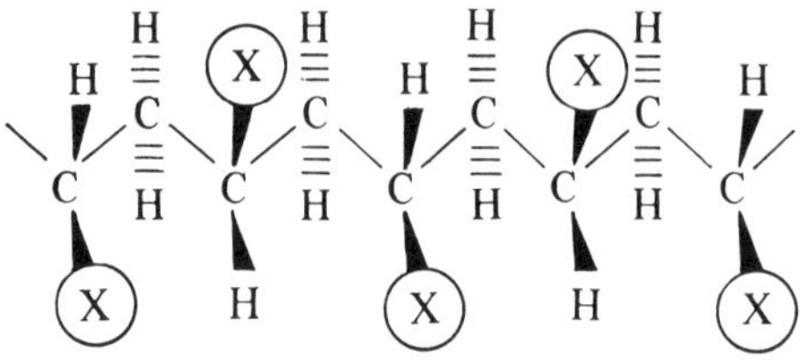

dldld

(c) *Atactic.* When the stereochemistry of the tertiary carbons in the chain is random the polymer is said to be atactic.

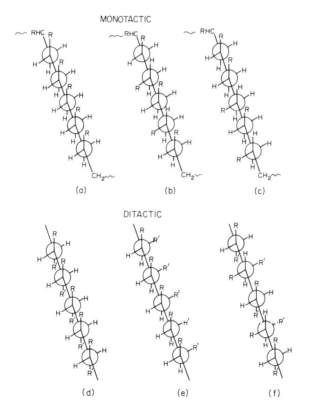

ddldll

It is often easier to obtain a clear picture of the spatial arrangement of the chains by referring to a Newman projection, and these are shown for comparison in figure 6.1a.

FIGURE 6.1. Newman projections of various stereoregular forms, (a) Isotactic, (b) Syndiotactic, (c) Atactic, (d) Erythro-di-isotactic, (e) Threo-di-isotactic, and (f) Di-syndiotactic.

DITACTIC POLYMERS

A more complicated picture emerges when the polymerization of 1,2-disubstituted ethylenes (CHR=CHR') is considered, because now each carbon atom in the chain becomes a chiral centre. The resulting ditactic structures are illustrated in figure 6.1b. Two isotactic structures are obtained, the *erythro,* where all the carbon atoms have the same configuration and the *threo,* in which the configuration alternates. Only one di-syndiotactic structure is possible. The differences arise from the stereochemistry of the starting material; if the monomer is *cis*-substituted the *threo* form is obtained whereas a *trans* monomer leads to the *erythro* structure.

POLYETHERS

When the spacing between the asymmetric centres increases, as in the hetero-atomic polymer poly(propylene oxide), the isotactic and syndiotactic structures become less easily recognized in planar projection. Using an extended zig-zag structure, the *isotactic* form now has its substituents alternating across the plane containing the main chain bonds.

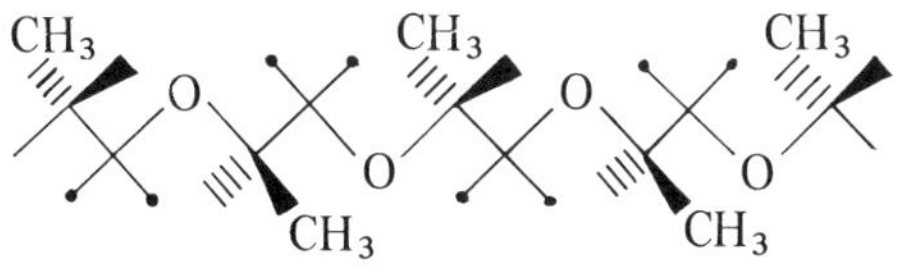

The reverse is true for the *syndiotactic* chain where the substituents are all located on one side.

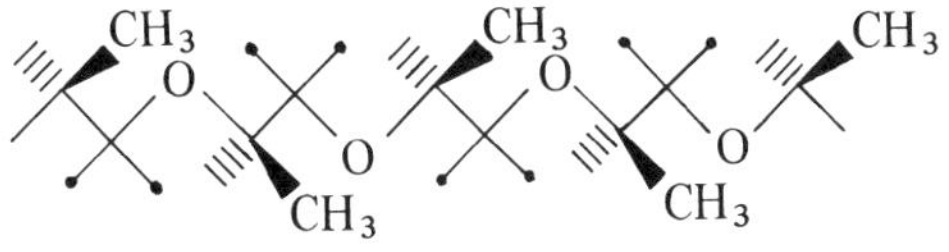

6.4 Geometric isomerism

In addition to the configurational isomerism encountered in polymers derived from asymmetric olefins, geometric isomerism is obtained when conjugated dienes are polymerized, *e.g.* $(CH_2=CX-CH=CH_2)$. Chain growth from monomers of this type can proceed in a number of ways, illustrated conveniently by 2-methyl-1,3-butadiene (isoprene). Addition can take place either through a 1,2-mechanism or a 3,4-mechanism, both of which could lead to isotactic, syndiotactic, or atactic structures, or by a 1,4-mode leaving the site of unsaturation in the chain.

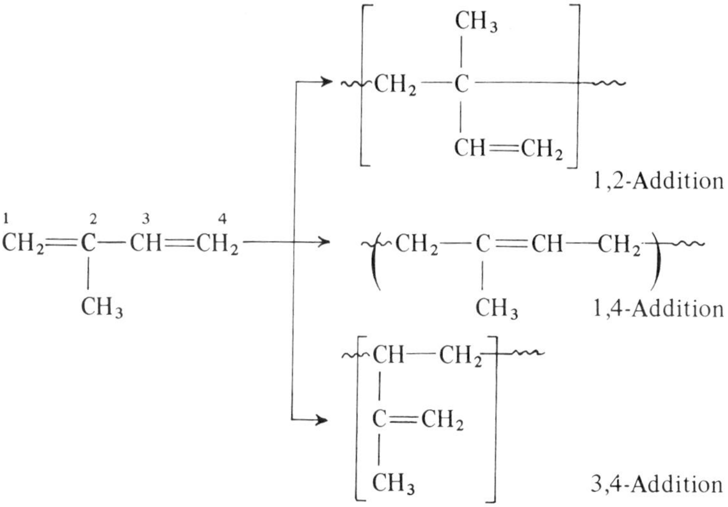

$$\begin{array}{c} CH_3 \\ | \\ \sim\!\!CH_2\!-\!C\!-\!\!\sim\!\!\sim \\ | \\ CH\!=\!CH_2 \end{array}$$

1,2-Addition

$$\begin{array}{cccc} 1 & 2 & 3 & 4 \\ CH_2\!\!=\!\!C\!-\!CH\!\!=\!\!CH_2 \\ | \\ CH_3 \end{array} \longrightarrow$$

$$\sim\!\!\left(\!\!CH_2\!-\!C\!\!=\!\!CH\!-\!CH_2\!\!\right)\!\!\!\sim$$
$$\begin{array}{c} | \\ CH_3 \end{array}$$

1,4-Addition

$$\begin{array}{c} \sim\!\!CH\!-\!CH_2\!\!\sim\!\!\sim \\ | \\ C\!\!=\!\!CH_2 \\ | \\ CH_3 \end{array}$$

3,4-Addition

This means that the 1,4 polymer can exist in the *cis* or *trans* form or a mixture of both.

$$\begin{bmatrix} \sim\!CH_2 & CH_2\!\!\sim \\ & \diagdown\!\!C\!\!=\!\!C\!\!\diagup \\ CH_3\!\!\diagup & \diagdown\!H \end{bmatrix}_n$$

$$\begin{bmatrix} \sim\!CH_2 & H \\ & \diagdown\!\!C\!\!=\!\!C\!\!\diagup \\ CH_3\!\!\diagup & \diagdown\!CH_2\!\!\sim \end{bmatrix}_n$$

cis   *trans*

In theory it is possible to synthesize eight distinguishable stereochemical forms or mixtures of these. For a symmetrical monomer such as 1,3-butadiene ($CH_2\!\!=\!\!CH\!-\!CH\!\!=\!\!CH_2$) the 1,2- and 3,4-additions are indistinguishable and the possible number of stereoforms diminishes accordingly.

Additional variations are possible when 1,4-disubstituted dienes are considered, $XCH\!\!=\!\!CH\!-\!CH\!\!=\!\!CHY$. For $Y = H$

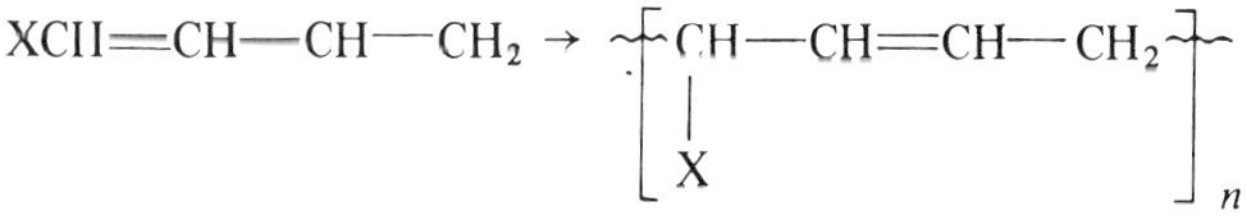

$$XCH\!\!=\!\!CH\!-\!CH\!-\!CH_2 \rightarrow \sim\!\!\left[\!\!\begin{array}{c} CH\!-\!CH\!\!=\!\!CH\!-\!CH_2 \\ | \\ X \end{array}\!\!\right]_n\!\!\sim$$

both *cis* and *trans* isomerism is possible together with isotactic, syndiotactic, and atactic placements for the group X when the addition is 1,4. When $Y \neq H$ then *threo* and *erythro* forms are also possible in the 1,4-polymer. The name *tritactic* has been suggested for polymers prepared from monomers with different X and Y groups.

6.5 Factors influencing stereoregulation

The low pressure polymerization of ethylene, reported by Ziegler in 1955, signalled the emergence of a new phase in polymer science. The catalyst, prepared from $TiCl_4$ and $(C_2H_5)_3Al$, was heterogeneous and it was subsequently demonstrated by Natta that co-ordination catalysts of this type could be used to exercise control over the stereoregular structure of the polymer. Initially it was thought that only heterogeneous catalysis would lead to stereoregular polymers, but we now know that this is untrue and that stereoregulation can be effected under specific rigorously defined conditions, regardless of the solubility of the catalyst system.

If stereoregulation is simply the control of the mode of entry of a monomer unit to a growing chain, examination of the factors influencing this addition should provide an understanding of how to exert such control.

Free radical initiation can usually be thought of as generating a chain by a Bernoulli-trial propagation, in which the orientation of the incoming monomer is unaffected by the stereostructure of the polymer. It can then add on in

one of two ways where the active end is assumed to be a planar sp^2 hybrid, and the configuration of the adding monomer is finally determined only when another monomer adds on to it in the next step. In other words this addition leads to an isotactic or syndiotactic placement of the pseudo-asymmetric centre 1 with respect to 2. When the chain carrier is a free species, *i.e.* a radical, the stereoregularity of the polymer is a function of the relative rates of the two. methods of addition and this is governed by the temperature. Consideration of the relative magnitudes of the enthalpy and entropy of activation for isotactic and syndiotactic placements shows that while the differences are small, the syndiotactic structure is favoured. This is, of course, aided by the greater steric hindrance and repulsions experienced by the substituents in the isotactic configuration and will vary in extent with the nature of the group X. Thus for a free radical polymerization at 373 K, the fraction of syndiotactic placements is 0.73 for methyl methacrylate monomer but only 0.51 when vinyl chloride is used. A decrease in the polymerization temperature increases the tendency towards syndiotactic placements, but as radical reactions are normally high temperature processes atactic structures predominate. Low temperature free radical propagation has been found to produce syndiotactic polymers from the polar monomers, isopropyl and cyclohexyl acrylate, and methyl methacrylate.

The same general principles apply for freely propagating ionic chain carriers, but if co-ordination between the monomer and the active end takes place, the stereoregulation is altered. The configuration of the monomer is then influenced by the stereochemistry of the growing end, and the possible number of ways the monomer can join the chain is in excess of two. These co-ordination catalysts include the Ziegler-Natta type as the largest group, and others such as butyl lithium, phenyl magnesium bromide, and boron trifluoride etherate. The resulting polymer is normally isotactic, although some cases exist where highly syndiotactic polymers are obtained.

TABLE 6.1. Polymerizations using co-ordination catalysts where quoted tacticities are $>$ 90 per cent

Monomer	Catalyst	Structure
Isobutyl vinyl ether	$BF_3(C_2H_5)_2O$ in propane at 213 K	Isotactic
Methyl acrylate	C_6H_5MgBr or $n\text{-}C_4H_9Li$ in toluene at 253 K	Isotactic
Propylene	$TiCl_4 + (C_2H_5)_3Al$ in heptane at 323 K	Isotactic
Propylene	$VCl_4 + Al(i\text{-}C_4H_9)_2Cl$ in anisole or toluene at 195 K	Syndiotactic

The orienting stage in co-ordination polymerization can be pictured as being multicentred, with the monomer position governed by co-ordination with the gegen-ion and the propagating chain end. As the gegen-ion will tend to repel the substituent X on the incoming monomer, it is forced to approach in a way that leads to predominantly isotactic placement. If co-ordination plays a major role in determining the configuration of the incoming monomer, then the greater the co-ordinating power the more regular the resulting polymer should be, but the nature of the monomer is also important. Polar monomers (the acrylates and vinyl ethers) are capable of taking an active part in the co-ordination process, and will only require catalysts with moderate powers of orientation, but non-polar monomers such as the α-olefins will require stronger co-ordinating catalysts to maintain the required degree of stereoregulation in the addition process. In extreme cases the heterogeneous Ziegler-Natta catalysts are required, where severe restrictions are imposed on the method of monomer approach to the growing chain end and these must be used for the non-polar monomers which yield only atactic polymers with homogeneous catalyst systems.

6.6 Homogeneous stereospecific cationic polymerizations

An example of this type of reaction is provided by the alkyl vinyl ethers $(CH_2{=}CHOR)$. Isobutyl vinyl ether was the first monomer studied which produced a stereoregular polymer using a $BF_3 + (C_2H_5)_2O$ catalyst and will be used as the illustrative monomer. A homogeneous stereospecific polymerization can be carried out in toluene at 195 K using such soluble complexes as $(C_2H_5)_2TiCl_2AlCl_2$ or $(C_2H_5)_2TiCl_2Al(C_2H_5)Cl$, or, if a suitable choice of mixed solvents is made, a homogeneous system with $BF_3 + (C_2H_5)_2O$ can be obtained which is capable of producing the isotactic polymer.

The mechanism proposed by Bawn and Ledwith, postulates the existence of an sp^3 configuration for the terminal carbon in the growing chain due to the attendant gegen-ion, and especially in low dielectric solvents. They also point out that the structure of the aklyl vinyl ethers, with the exception of the ethyl and isopropyl members, will be subject to steric shielding of one side of the double bond, i.e.

This blocks one mode of double-bond opening and assists stereoregulation. This conclusion is supported by the lack of any crystalline polymer in the product when the ethyl and isopropyl groups are used, where no blocking is possible.

The formation of a loose six-membered ring is thought to stabilize the growing carbonium ion in the reaction so that the only route for monomer approach is past the counter-ion.

Isotactic polymer

A four-centred cyclic transition state is involved in the propagation stage leading to the insertion of a monomer unit between the catalyst and the chain end, with subsequent regeneration of the cyclic structure. An alternative transition state, proposed by Cram and Kopecky, has a similar but more rigid structure.

Both mechanisms ignore the nature of the catalyst forming the gegen-ion, but obviously as this will act as a template for the attacking monomer it will exert an influence on the rate of reaction and the type of stereoregularity imposed. The most probable configuration is isotactic because of the tendency for the gegen-ion to repel the substituent group of the incoming monomer.

6.7 Homogeneous stereoselective anionic polymerizations
The various factors influencing the stereoregularity, when the propagating chain end is a carbanion, are conveniently highlighted in a study of the polymerization of methyl methacrylate by organo-lithium catalysts.

The propagating chain end in an anionic reaction initiated by a reagent such as n-butyl lithium can be thought of as existing in one of the following states, analogous to carbonium ion formation.

$$RLi \; \rightleftharpoons \; R^-Li^+ \; \rightleftharpoons \; R^-//Li^+ \; \rightleftharpoons \; R^- + Li^+$$

covalent   contact ion   solvent separated   free ions
                pair   ion pair

The extent of the separation will depend on the polarity of the reaction medium and in non-polar hydrocarbon solvents, such as toluene, covalent molecules or contact ion pairs, are most likely to exist. With increasing solvent polarity there is a greater tendency to solvate the ions, eventually producing free ions for strictly anionic polymerizations. These lead to conditions similar to a free radical , polymerization where the stereoregulation is reduced and syndiotactic placements are favoured at low temperatures.

The effects of solvent and temperature are manifest in the polymerization of methyl methacrylate with n-butyl lithium at 243 K in a series of mixed solvents prepared from toluene and dimethoxyethane (DME). The n.m.r. spectra of the products indicate the compositions in table 6.2 and reveal that a predominantly isotactic material is produced in a low polarity medium, but that this becomes highly syndiotactic as the solvating power of the medium increases.

TABLE 6.2. The effect of mixed solvent composition on the tacticity of poly(methyl methacrylate) initiated by n-butyl lithium at 243 K; the mole fractions of the various configurations are given

Toluene/DME	Isotactic	Heterotactic	Syndiotactic
100/0	0.59	0.23	0.18
64/36	0.38	0.27	0.35
38/62	0.24	0.32	0.44
2/98	0.16	0.29	0.55
0/100*	0.07	0.24	0.69

* measured at 203 K.

An additional point emerges from this; higher syndiotactic contents are obtained when the Lewis base strength of the solvent increases and this factor probably accounts for the efficiency of the ether in this system. When the catalyst is 9-fluorenyl lithium, the reaction of methyl methacrylate at 195 K in toluene leads to isotactic polymer, whilst a change of solvent to tetrahydrofuran results in a syndiotactic product.

Stereoregulation is also altered by the nature of the solvent when Grignard reagents and alkali metal alkyls are used as initiators. In toluene, for example, the isotactic placements in the chain decrease as reagents change from Li to Na to K.

If general conclusions can be drawn from the behaviour of methyl metha-crylate, it appears that stereoregulation in anionic polymerizations, involving either polar monomers or monomers with bulky substituents, will lead to predominantly syndiotactic polymers when a free, dissociated ion, occurs at the propagating end. This is because it is the most stable form arising from a minimization of steric and repulsive forces. If, however, some strongly regulating kinetic mechanism is available, for instance monomer + gegen-ion co-ordination, · then the less favourable isotactic placement occurs.

In an attempt to explain the mechanism of stereoregulation in systems catalysed by lithium alkyls, Bawn and Ledwith proposed that the penultimate residue plays an important role in the addition. A loose cyclic intermediate is formed when the Li^+ counter-ion co-ordinates with the carbonyl of the penultimate unit and with the terminal unit in a resonance enolic structure V.

V   VI

This can be represented alternatively as a transition state VI similar to that in an S_N2 reaction. With one side of the Li^+ shielded, monomer approach is restricted, and the path of least resistance places the α-methyl group on the incoming monomer in a *trans* position, relative to the α-methyl group on the carbanion, during the π-complex formation. Addition then proceeds through a series of bond exchanges as the carbanion joins the monomermethylene group. The carbonyl group of the monomer co-ordinates with the ion by replacing the interaction with the previous penultimate group and the cyclic intermediate is regenerated.

The steric restriction imposed by the α-methyl group aids the formation of an isotactic polymer and in its absence (*i.e.* methyl acrylate) there is a reduced probability of isotactic placements. A compensating feature in the higher acrylates arises from the shielding of one side of the monomer by the bulkiness of the ester group. In the branched homologues, isopropyl and *t*-butyl acrylate, π-bonding with the Li^+ ion is forced to take place on one side of the monomer only, thereby enhancing the formation of isotactic polymer quite markedly.

As this and other mechanisms all postulate the existence of structures stabilized by intramolecular solvation, the addition of Lewis bases or polar solvents should disrupt the required template and encourage conventional anionic propagation by free ions. This automatically reduces the probability of an isotactic placement occurring.

6.8 Homogeneous diene polymerization

The principles applied in the previous section to essentially polar monomers can be extended to the stereoregular polymerization of dienes by alkali metals and metal alkyls. We have already seen that the *cis-trans* isomerism presents a variety of possible structures for the polydiene to adopt and complicates the preparation of a sample containing only one form rather than a mixture. Thus polyisoprene may contain units in the 1,2 or 3,4 or *cis*-1,4 or *trans*-1,4 configuration without even considering the tacticity of the 1,2 or 3,4 monomer sequences in the chain.

Most work has centred on the preparation of a particular form of geometric isomer because the type and distribution of each isomeric form in the chain has a profound influence on the mechanical and physical properties of the sample. The original discovery that metallic lithium in a hydrocarbon solvent catalysed the production of an all *cis*-1,4 polyisoprene stimulated interest in this area and quickly raised two points which must be satisfied if a suitable mechanism is to be postulated.

(i) Lithium and lithium alkyl catalysts produce highly specific stereostructures, but when replaced by Na or K this effect diminishes.
(ii) Stereospecific polymerization takes place in the bulk state or in hydrocarbon solvents, but the addition of a polar solvent leads to drastic changes.

To explain these features the following mechanism has been put forward. Initiation produces a "Schlenk" adduct VII.

cis-1,4

The lithium ion then forms a chelate complex with the isoprene monomer locking it into a *cis*-configuration which is maintained during the addition reaction. This type of complex is suitable when a small ion like Li^+ is used but will be disrupted by the larger gegen-ions Na^+ and K^+, thereby allowing freer approach of the reactants. The presence of ethers also alters the stereospecificity by competing for the Li^+ and altering the spatial arrangement of the chelating pattern.

The monomer can then enter in a random fashion. The absence of significant 1,2- or 3,4-addition is thought to be caused by the shielding of carbon 3 in the transition state. However, all such proposals remain speculative.

6.9 Summary

We can now summarize a few important points dealt with so far.

Three factors influence stereoregularity during chain propagation:

(1) *Steric factors* which force the unit into a spatial arrangement determined by the size and position of the substituents already in the chain;

(2) *Polar factors* because solvents which allow contact ion pairs favour isotactic placements, but pairs separated by heavy solvation (free ions) lead to syndiotactic structures;

(3) *Co-ordination* because if the end group of a growing chain has a planar (sp^2) configuration with no established parity, such as that found in free radical or free ion propagation, then the configuration of this unit is established only

during the course of addition of an incoming monomer. Normally this will result in a syndiotactic placement with respect to the penultimate unit. Otherwise, co-ordination occurs between the gegen-ion, the incoming polar monomer, and the end or penultimate unit.

For polar monomers the soluble catalysts can produce isotactic structures but for non-polar monomers the homogeneous catalysts lead mainly to atactic or syndiotactic polymers and a heterogeneous catalyst is required for isotactic placements to occur. These will now be discussed.

6.10 Polymerization using Ziegler-Natta catalysts

Stereoregular polymerizations carried out in homogeneous systems, using essentially polar monomers whose ability to co-ordinate with the catalyst-complex imposes a stereospecific mechanism on the addition, have been dealt with above. As the polarity of the monomer decreases, however, the ability to control the configuration of the incoming monomer decreases and atactic polymers result.

The work carried out by Ziegler on organometallic catalyst systems, culminating in the discovery that ethylene can be polymerized at ambient temperatures and atmospheric pressure, opened the way to the preparation of stereoregular α-olefins. Ziegler found that the polyethylene obtained, using a catalyst formed by mixing solutions of triethyl aluminium and titanium tetrachloride, was a highly linear crystalline polymer, as opposed to the branched more amorphous material obtained at high pressures. This stimulated Natta to further work, and he and his co-workers demonstrated that highly crystalline linear polymers from propylene, 1-butene, and a number of other α-olefins, could be prepared using modified catalysts of the Ziegler type. It was also found that the crystallinity in these polymers arose from their highly stereoregular structure.

The systems were in all cases heterogeneous and the active initiators are now known by the general name *Ziegler-Natta* catalysts. This encompasses a vast number of substances prepared from different combinations of organometallic compounds where the metal comes from the main Groups I, II, or III and is combined with the halide or ester of a transition metal (Groups IV to VIII). Table 6.3 contains a number of the common components of the Ziegler-Natta catalysts but this list is far from exhaustive.

These catalysts tend to control two features, (a) the rate and (b) the specificity of the reaction, but this varies from reaction to reaction and only a judicious choice of catalyst can effect control over both of these aspects.

Unfortunately the insolubility of the catalyst poses the problems that the kinetics are hard to reproduce and the reaction mechanisms are difficult to formulate with real confidence. This means that the choice of a suitable catalyst for a system is somewhat empirical and very much trial and error, until optimum conditions are established.

It is useful to remember that both heterogeneous and homogeneous catalysts exist in the Ziegler-Natta group but the latter only yield atactic or occasionally

TABLE 6.3. Components of Ziegler-Natta Catalysts

Metal Alkyl or Aryl	Transition metal compounds
$(C_2H_5)_3Al$	$TiCl_4$; $TiBr_3$
$(C_2H_5)_2AlCl$	$TiCl_3$; VCl_3
$(C_2H_5)AlCl_2$	VCl_4; $(C_5H_5)_2TiCl_2$
$(i\text{-}C_4H_9)_3Al$	$(CH_3COCHCOCH_3)_3V$
$(C_2H_5)_2Be$	$Ti(OC_4H_9)_4$
$(C_2H_5)_2Mg$	$Ti(OH)_4$; $VOCl_3$
$(C_4H_9)Li$	$MoCl_5$; $CrCl_3$
$(C_2H_5)_2Zn$	$ZrCl_4$
$(C_2H_5)_4Pb$	$CuCl$
$((C_6H_5)_2N)_3Al$	WCl_6
C_6H_5MgBr	$MnCl_2$
$(C_2H_5)_4AlLi$	NiO

syndiotactic polymers from non-polar monomers. As only the heterogeneous Ziegler-Natta catalysts produce isotactic poly-α-olefins, these have received most attention. The interest in this type of system has been immense, as evidenced by the vast quantity of published material, and it was most fitting that both Ziegler and Natta were recognized for their work by being awarded jointly the Nobel Prize for chemistry in 1963.

EXPERIMENTAL DEMONSTRATION

Before dealing with the mechanism of catalysis and the nature of the catalyst in greater depth, a description of a laboratory preparation of polyethylene will be given to demonstrate the use of such systems.

Preparation of the catalyst. This is prepared from either aluminium triethyl or aluminium diethyl chloride in combination with titanium tetrachloride. The main disadvantage in using aluminium alkyls is that they ignite spontaneously in air and, to avoid this hazard, must be handled in an inert atmosphere.

A safer procedure is to use amyl lithium which can be prepared from lithium wire and amyl chloride. Petroleum ether ($50\ cm^3$) is stirred in a three-necked flask and degassed under a stream of nitrogen, which is first purified by passing through a pyrogallol and sodium hydroxide train to remove oxygen. Lithium wire (3 g) is added, followed by $2\ cm^3$ of a solution of amyl chloride ($20.7\ cm^3$) in petroleum ether ($25\ cm^3$). This is stirred vigorously until the solution becomes turbid (LiCl) and then the remaining amyl chloride solution is added slowly over a period of 20 min to the reaction flask now being cooled in an ice-bath. The reaction mixture turns blue-brown and after 2.5 h it is filtered under nitrogen through glass wool into a graduated flask to remove unreacted Li. The filtrate is allowed to settle and the supernatant liquid is analysed by hydrolysing an aliquot with water and titrating the LiOH formed with 0.1 M HCl. The amyl lithium solution can be stored for some days at 273 K.

Reaction. The catalyst is prepared *in situ.* An apparatus similar to that in figure 6.2 is used. The reaction kettle ($1 \ dm^3$) is charged with $400 \ cm^3$ petroleum ether and 0.05 mol lithium amyl. Anhydrous $TiCl_4$ ($2 \ cm^3$) is added and the formation of the catalyst as a brownish-black precipitate is complete in 20 min. The formation is accompanied by a rise in temperature of about 10 K. Ethylene is then passed into the stirred mixture and polyethylene forms immediately. The reaction can be allowed to continue for 30 min, then the catalyst is destroyed by the addition of butanol ($40 \ cm^3$). The polymer is filtered, washed with a 1:1 mixture of HCl and methanol, and dried at 350 K. It has a high degree of crystallinity, a higher density, and a melting temperature some 20 to 30 K higher than samples prepared using high pressure techniques.

The apparatus shows a syringe in position which may be used if the more inflammable aluminium alkyl catalysts are used, as these are often handled in hydrocarbon solvents.

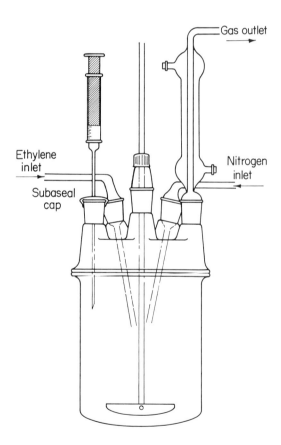

FIGURE 6.2. Apparatus for polymerization of ethylene.

6.11 Nature of the catalyst

Frequently the product of a Ziegler-Natta polymerization is sterically impure and can be preferentially extracted to give two products – a highly crystalline stereoregular fraction, and an amorphous atactic one. This may be attributed to the size of the catalyst particles as stereoregularity is enhanced by having large particles, whereas a finely divided catalyst tends to produce an amorphous polymer.

The crystal form of the catalyst is also important and the violet α, γ, and δ forms of $TiCl_3$ produce a greater quantity of isotactic polypropylene when combined with an aluminium alkyl, than the brown β-structure. As the active sites for heterogeneous polymerizations are believed to be situated on the crystal surfaces, the structure is all important. In the layered structure of α-$TiCl_3$, where every third Ti^{3+} ion in the lattice is missing, a number of Cl vacancies occur on the surface to maintain electrical neutrality in the crystal. The Ti^{3+} on the surface is then only 5-co-ordinated leaving a vacant d-orbital, $\square$, and an active site is created when an alkyl group replaces a chloride ion to form $TiRCl_4$ $\square$. (See diagrams page 123.)

In β-$TiCl_3$, the linear chains form bundles in which some Ti ions are surrounded by five Cl^- ions and some by only four Cl^-. This means that the steric control at the sites with two vacancies is now less rigid and stereoregulation is much poorer.

Catalyst composition affects both stereoregulation and polymer yield. Thus Ti^{3+} is a more active producer of isotactic polypropylene than Ti^{4+} or Ti^{2+}, while an increase in the length of the associating alkyl group decreases the efficiency of stereoregular placements. Varying the transition metal and the associated aluminium compounds in the catalysts also influences the nature of the product.

6.12 Nature of active centres

Most of the experimental evidence points to propagation taking place at a carbon to transition metal bond with the active centre being anionic in character. Free radical reactions are considered to be non-existent in the Ziegler-Natta systems because neither (i) chain transfer nor (ii) catalyst consumption occurs. The active centres also live longer than radicals and resemble "living" polymer systems in many ways, one being that block copolymers can be produced by feeding two monomers alternatively into the system.

While a number of reaction mechanisms have been suggested, two are worth considering in detail. These are based on the view that the active centres are localized rather than migrating and that the α-olefin is complexed at the transition metal centre prior to incorporation into the chain, *i.e.* growth is always from the metal end of the growing chain.

The active species are then considered to be either *bimetallic* or *monometallic*.

6.13 Bimetallic mechanism

Natta and his associates have postulated a mechanism involving chain propagation from an active centre formed by the chemisorption of an electropositive metal alkyl of small ionic radius on the co-catalyst surface. This yields an electron deficient bridge complex such as **VIII** and chain growth then emanates from the C–Al bond.

$$\text{Cl}\diagdown\underset{\text{Cl}\diagup}{\text{Ti}}\cdots\underset{\cdot\cdot\text{Cl}\cdot\cdot}{\overset{\diagup\text{R}\diagdown}{}}\cdots\underset{\text{R}}{\overset{\diagup\text{R}}{\text{Al}}}$$

VIII

It is suggested that the nucleophilic olefin forms a π-complex with the ion of the transition metal and, following a partial ionization of the alkyl bridge, the monomer is included in a six-membered ring transition state. The monomer is then incorporated into the growing chain between the Al and the C allowing regeneration of the complex.

While a limited amount of experimental evidence does lend support to this concept, major objections have been voiced by Ziegler, who is of the opinion that as dimeric aluminium alkyls are inefficient catalysts in the "Aufbau" reaction, the Ti–Al complex is not likely to be the effective catalytic agent. Other more recent work also favours the second and simpler alternative, the monometallic mechanism.

6.14 Monometallic mechanism

Majority opinion now favours the concept that the d-orbitals in the transition element are the main source of catalytic activity and that chain growth occurs at the titanium-alkyl bond. The ideas now presented are predominantly those of Cossee and Arlman, and will be developed using propylene as the monomer.

The first stage is the formation of the active centre illustrated here using
α-$TiCl_3$ as catalyst. The suggestion is that alkylation of the 5-co-ordinated Ti^{3+}

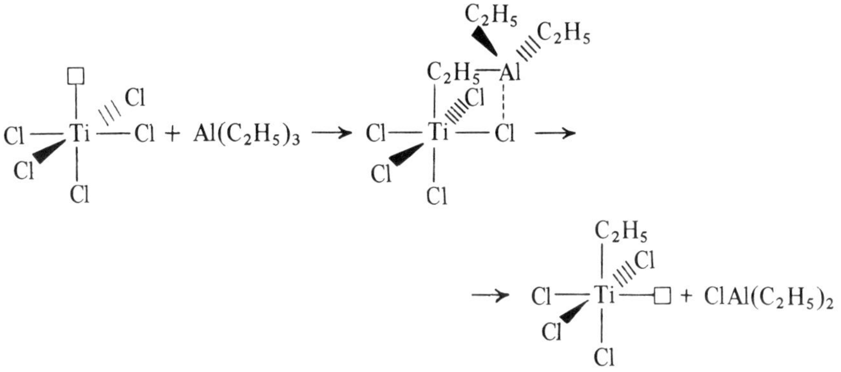

$$Cl-Ti \overset{Cl}{\underset{Cl}{\big|}} Cl + Al(C_2H_5)_3 \longrightarrow Cl-Ti \overset{C_2H_5-Al}{\underset{Cl}{\big|}} Cl \longrightarrow$$

$$\longrightarrow Cl-Ti \overset{C_2H_5}{\underset{Cl}{\big|}} \square + ClAl(C_2H_5)_2$$

ion takes place by an exchange mechanism after chemisorption of the aluminium
alkyl on the surface of the $TiCl_3$ crystal. The four chloride ions remaining are
the ones firmly embedded on the lattice and the vacant site is now ready to
accommodate the incoming monomer unit. The reaction is confined to the
crystal surface and the active complex is purely a surface phenomenon in
heterogeneous systems.

The attacking monomer is essentially non-polar but forms a π-complex with
the titanium at the vacant d-orbital. A diagram of a section of the complex

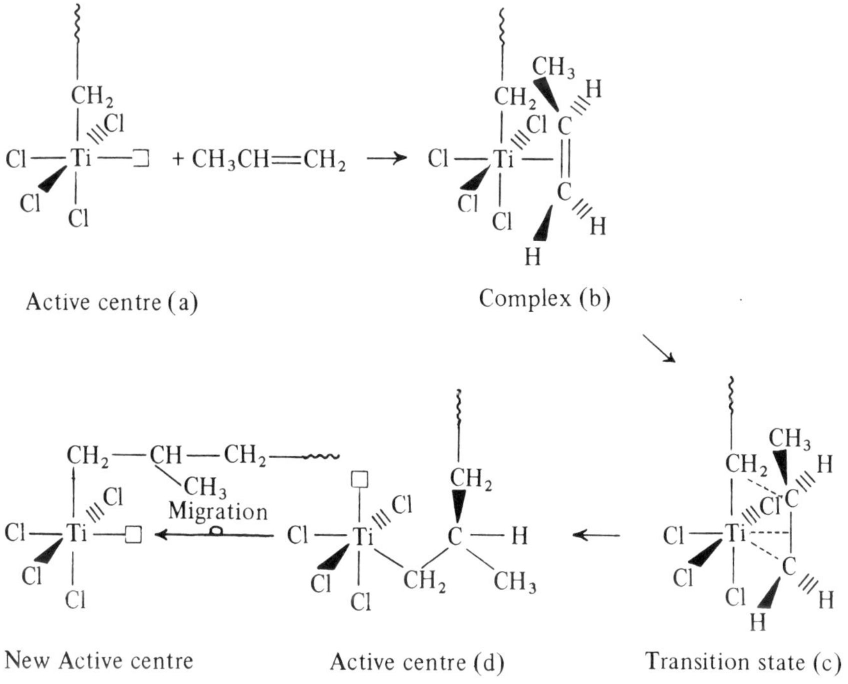

$$Cl-Ti \overset{CH_2}{\underset{Cl}{\big|}} \square + CH_3CH{=}CH_2 \longrightarrow Cl-Ti \overset{CH_2}{\underset{Cl}{\big|}} {-}\|{-}$$

Active centre (a)   Complex (b)

New Active centre   Active centre (d)   Transition state (c)

Migration

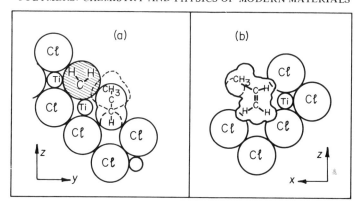

FIGURE 6.3. Cross-sectional diagrams of the propylene-catalyst complex through (a) the $y-z$ plane and (b) the $x-z$ plane of the octahedral structure. (Adapted from Bawn and Ledwith.)

shows that the propylene molecule is not much bigger than a chloride ion and consequently the double bond can be placed adjacent to the Ti ion and practically as close as the halide. After insertion of the monomer between the Ti–C bond, the polymer chain then migrates back into its original position ready for a further complexing reaction.

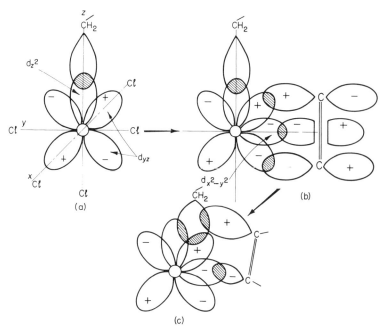

FIGURE 6.4. Representation of the relevant orbital overlap in (a) the active centre, (b) the titanium-olefin complex, and (c) the transition state.

The reactivity of the active centre is attributed primarily to the presence of d-orbitals in the transition metal. The initial state of the active centre shows that the $=CH_2$ group will be capable of considerable distortion from its equilibrium position, because of the availability of adjacent d-orbitals. Complexing (b) takes place when the π-bonding orbitals of the olefin overlap with the vacant $d_{x^2-y^2}$ orbital of the Ti^{3+} while at the same time the π^*-antibonding orbitals can overlap the d_{yz} orbitals of the Ti^{3+}. Formation of the transition state is aided by the ability of the $=CH_2$ group to migrate by partial overlap with the d_{z^2}, d_{yz}, and π^*-orbitals.

The main features of the monometallic mechanism are: (1) an octahedral vacancy on the Ti^{3+} is available to complex the olefin; (2) the presence of an alkyl to transition metal bond at this site is required; and (3) the growing polymer chain is always attached to the transition metal.

6.15 Stereoregulation

To obtain a stereoregular polymer, the chemisorption of the monomer on the catalyst surface must be controlled so that the orientation of the incoming monomer is always the same. Examination of models shows that a molecule such as propylene will fit into the catalyst surface in only one way if a position of closest approach of the double bond to the Ti^{3+} ion is to be achieved. This places the $=CH_2$ group pointing into the lattice and for steric reasons the orientation of the $-CH_3$ group to one side is preferred. This determines the configuration of the monomer during the complexing stage and is always the same. Repeated absorption of the monomer in this orientation, prior to reaction, leads to an isotactic polymer.

For the Cossee-Arlman mechanism to operate, migration of the vacant site back to its original position is necessary, else an alternating position is offered to the chemisorbed monomer and a syndiotactic polymer would result. This implies that the tacticity of the polymer formed depends essentially on the rates of both the alkyl shift and the migration. As both of these will slow down when the temperature is decreased, formation of syndiotactic polymer should be favoured at low temperatures, and syndiotactic polypropylene can in fact be obtained at 203 K.

6.16 Natural synthetic rubber – "Natsyn" (IR)

The original goal of most chemists interested in elastomers was the synthesis of *cis*-polyisoprene, natural synthetic rubber, or "Natsyn" as it is sometimes called. The problem was complicated by the ability of the starting material, isoprene, to polymerize in four different ways and form mixtures of the various combinations. With the advent of stereospecific Ziegler-Natta catalysts the "impossible" was achieved in 1956.

In the preparation of *cis*-polyisoprene, great care must be taken in the purification of reagents. Various catalysts are available, but the trialkylaluminium

+ titanium tetrachloride combination has been found to produce a *cis*-content in excess of 94 per cent.

A 15 per cent solution of high purity isoprene in dry pentane (99 per cent) is prepared and 55 g, drained directly from a cooled silica gel column, are added to a clean dry screw cap bottle. The contents of the bottle are reduced to about 50 g by heating the open bottle on a sand bath before adding by syringe 0.200 mmol of aluminium tri-isobutyl and 0.185 mmol of titanium tetrachloride. Both of these compounds react violently with oxygen or water and are conveniently handled by first preparing 0.2 to 0.5 mol dm^{-3} solutions in pure dry heptane which are stored under nitrogen in bottles sealed by serum caps. These can be kept in a dry-box in an inert atmosphere, and, when required, a quantity is withdrawn by syringe.

After the reagents have been added, the bottle is sealed with a teflon lined cap and rotated for 16 h in a thermostat bath at 323 K. When the reaction is complete and cool, an antioxidant is added (2 per cent solution of di-*tert*-amyl hydroquinone in benzene is suitable). The mixture is then poured into isopropyl alcohol (0.2 dm^3) containing 2 g of antioxidant. The precipitated polymer is separated and dried at 310 K under vacuum.

The *trans*-1,4-polyisoprene (99 per cent) can be prepared using an aluminium + titanium + vanadium catalyst. "Natsyn" has a very small share of world markets at present, but this may change as demands increase.

6.17 Conformation of stereoregular polymers

Many of the stereoregular polymers prepared are highly crystalline, and the tendency to form ordered structures increases as the stereoregularity becomes more pronounced. We shall see later that crystalline order is usually associated with regular symmetrical polymer structures, whereas the asymmetric monomers form highly unsymmetrical chains. Some other factors must aid crystallite formation.

A stable form of polyethylene is the all *trans* zig-zag form in which it crystallizes. An extended zig-zag pattern becomes untenable, however, for an isotactic polymer with a bulky substituent because the distance between the substituent centres in this conformation is only 0.254 nm. Obviously, the low energy form for an isotactic species must be attained by placing the substituents in staggered positions of maximum separation, and this is achieved when bond rotation generates a helix. One particular helical form is shown for polypropylene in figure 6.5. Working from carbon 1 we have the following sequence: 1 and 4 are *trans* to each other (*t*), carbons 2 and 5 are *gauche* (*g*), 3 and 6 are *trans*, 4 and 7 are *gauche*, and so on. Carbon 1 repeats at carbon 7, hence the helix is three fold with three monomer residues constituting one complete turn. In a shorthand notation this is a 3_1-helix with a *tgtgtg* conformation, and departure from this pattern to a *ttgg* sequence would simply lead the chain back on itself. This type of helix can also be built up on a triangular template which can be used as a simple model to demonstrate the structure. A helix generated in this

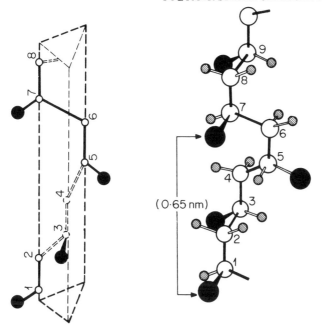

FIGURE 6.5. A 3₁-helix formed by a poly α-olefin, in this case polypropylene. This structure is also seen to fit a triangular template.

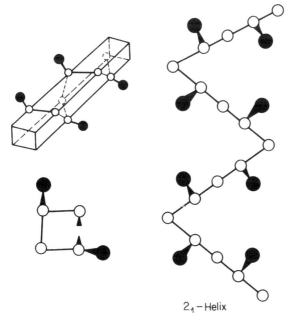

2₁ – Helix

FIGURE 6.6. A poly α-olefin in the syndiotactic configuration showing the *ttgg* sequence along the chain, and the two fold helix which fits a square template.

way using rotations of 120° should result in an identity period of 0.62 nm. Polypropylene has a period of 0.65 nm and the structure may be generated with equal ease either as a left- or right-handed helix. A helix turning in a clockwise direction, when the chain is viewed along its axis is said to be a right-handed helix; if anti-clockwise, it is left-handed.

The syndiotactic configuration is much more suitable for the extended zig-zag form as the substituents are already staggered for convenient packing on either side of the chain, but a two-fold helix can also be generated by adopting a *ttgg* sequence which has been identified in syndiotactic polypropylene. A square mandrel can be used to demonstrate this structure. The type of helix formed depends largely on the size of the substituent, and a number of these are shown in figure 6.7. As the helix is a regular ordered structure, it can be arranged compactly in a three-dimensional close structure with relative ease which explains how the unsymmetrical chain monomer can be accommodated in a

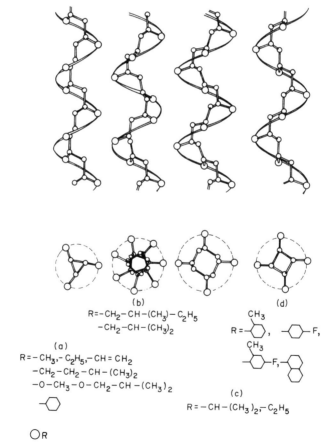

(b)
$R=-CH_2-CH-(CH_3)-C_2H_5$
$-CH_2-CH-(CH_3)_2$

(d)

(a)
$R=-CH_3,-C_2H_5,-CH=CH_2$
$-CH_2-CH_2-CH-(CH_3)_2$
$-O-CH_3-O-CH_2-CH-(CH_3)_2$

$R=-\langle\ \rangle CH_3$, $-\langle\ \rangle-F$,

$\langle\ \rangle-F$, $\langle\langle\ \rangle\rangle$
CH_3

(c)
$R=-CH-(CH_3)_2,-C_2H_5$

○R

FIGURE 6.7. A diagrammatic representation of various other ordered helical structures adopted by isotactic polymers. (After Natta and Corradini.)

crystalline polymer structure. Highly crystalline samples are obtained when the polymer is sufficiently stereoregular to enable it to form significantly long helical or regular zig-zag sections for ordered chain arrangement to take place.

The automatic identification of crystallinity with stereoregularity should be avoided, however, as they are not necessarily synonymous, and while highly stereoregular polymers tend to be crystalline, the existence of any polymer in a crystalline state does not automatically mean the sample is markedly stereoregular.

General Reading

W. Cooper, "Stereospecific polymerization", in *Progress in High Polymers,* Vol. I. Academic Press (1961).

M. Goodman, "Concepts of polymer stereochemistry", *Topics in Stereochemistry,* Vol. 2. Wiley-Interscience (1967).

A. D. Ketley, *The Stereochemistry of Macromolecules,* Vols. I–III. Edward Arnold (1968).

G. Natta, "Precisely constructed polymers", *Scientific American,* **205**, 33 (1961).

G. E. Schildknecht, "Stereoregular polymers" in *Encyclopaedia of Chemistry.* Reinhold Publishing Corp. (1966).

R. B. Seymour, *Introduction to Polymer Chemistry,* Chapter 6. McGraw-Hill (1971).

References

1. C. E. H. Bawn and A. Ledwith, *Quarterly Reviews,* **16**, 361 (1962).
2. G. Natta and P. Corradini, *Rubber Chem. Technol ,* **33**, 703 (1960).

CHAPTER 7

Polymers in Solution

7.1 Thermodynamics of polymer solutions

The interaction of long chain molecules with liquids is of considerable interest from both a practical and theoretical view point. For linear and branched polymers, liquids can usually be found which will dissolve the polymer completely to form a homogeneous solution, whereas cross-linked networks will only swell when in contact with compatible liquids. In this chapter we shall deal with linear or branched polymers and treat the swelling of networks in chapter 13.

When an amorphous polymer is mixed with a suitable solvent, it disperses in the solvent and behaves as though it too is a liquid. In a good solvent, classed as one which is highly compatible with the polymer, the liquid-polymer interactions expand the polymer coil, from its unperturbed dimensions, in proportion to the extent of these interactions. In a "poor" solvent, the interactions are fewer and coil expansion or perturbation is restricted.

The fundamental thermodynamic equation used to describe these systems relates the Gibbs free energy function G to the enthalpy H and entropy S, *i.e.* $G = H - TS$. A homogeneous solution is obtained when the Gibbs free energy of mixing $\Delta G^m \leqslant 0$, *i.e.* when the Gibbs free energy of the solution G_{12} is lower than the Gibbs functions of the components of the mixture G_1 and G_2.

$$\Delta G^m = G_{12} - (G_1 + G_2) \tag{7.1}$$

To understand the behaviour of polymers in solution more fully, however, a knowledge of the enthalpic and entropic contributions to ΔG^m is essential. The thermodynamic conditions which typify most non-polar polymer solutions, expressed in the partial molar quantities, are $\Delta H_1 \geqslant 0$, $\Delta S_1 \gg - R \ln x_1$ and these deviate markedly from the ideal solution for which $\Delta H_1 = 0$ and $\Delta S_1 = - R \ln x_1$. Here 1 denotes the solvent and x its mole fraction. For good solvents, these deviations are large and ΔG^m is negative, but as the solvent becomes poorer the deviations decrease until pseudo-ideal or "theta" conditions are eventually attained when $\Delta H_1 = T\Delta S_1$. When this occurs the polymer coil is in its unperturbed state and this is important for polymers.

Deviations from ideality embodied in ΔH^m are caused by the need to break solvent $(1-1)$ contacts and polymer $(2-2)$ contacts to form new polymer-solution $(1-2)$ contacts, which for unlike molecules with different cohesive energies usually leads to a positive ΔH^m. This can be expressed as

$$\Delta H^m = z\Delta e N_1 \phi_2, \qquad (7.2)$$

where N_1 is the number of solvent molecules, z is a lattice co-ordination number, ϕ_2 is the volume fraction of the polymer and Δe is the energy of formation of a $(1-2)$ contact.

The large deviations in ΔS^m can be attributed to the great differences in size between the polymer and solvent molecules. The movement of an individual polymer chain in the solid polymer is severely impeded by neighbouring chains, but on dissolution most of these restrictions are removed and the chain, which is in continual motion, can now adopt a vastly increased number of equi-energetic conformations. This results in a conformational entropy contribution in addition to that obtained from simple mixing and is largely responsible for the non-ideal ΔS^m.

7.2 Flory-Huggins theory

The deviations from ideality can be used to measure the compatibility of a polymer with a liquid and the first significant attempts to treat this theoretically were made by Flory and Huggins. Both chose a simple lattice representation for the polymer solution and calculated the entropy change on a statistical basis by estimating the total number of ways the polymer and solvent molecules could be arranged on the lattice. This was effected on the assumption that the size of a polymer segment was comparable to that of a solvent molecule.

The final expression for the partial molar Gibbs free energy of dilution was

$$(\partial \Delta G^m / \partial n_1) = \Delta G_1 = RT\{ln(1-\phi_2) + (1-1/x_n)\phi_2 + \chi_1 \phi_2^2\} . \qquad (7.3)$$

Here the dimensionless parameter $\chi_1 = z\Delta e/RT$ is known as the Flory-Huggins interaction parameter, n_1 is the amount of solvent, and $x_n = (V_2/V_1)$ is the number average degree of polymerization for a heterogeneous polymer. The first two terms in the brackets represent the entropic contribution and the final term the enthalpic contribution.

Unfortunately the simple lattice treatment does not describe the behaviour of dilute polymer solutions particularly well. The following simplifications in the theoretical treatment are invalid:
(1) a uniform density of lattice site occupation was assumed, but this only holds for concentrated solutions; (2) the assumption was made that the segment locating process is purely statistical, but this would only be true if Δe was zero, and (3) the treatment assumed that the flexibility of a chain is unaltered on passing into

solution from the solid state. This limits the calculation of the entropy of mixing to the combinatorial or geometric contribution only (*i.e.* a measure of the total number of molecular arrangements on the lattice) and neglects the contribution from the continual flexing of the chain − the non-combinatorial entropy.

Attempts to modify this theory, by recognizing that χ_1 is actually a free energy parameter composed of entropic χ_S and enthalpic χ_H contributions, have been made by Huggins, Miller, and Guggenheim. The expression for χ_1 then becomes

$$\chi_1 = \chi_H + \chi_S,$$

where $\qquad \chi_H = -T(\partial \chi_1 / \partial T)$ and $\chi_S = \partial(T\chi_1)/\partial T$.

Regrettably these refinements are still insufficient to explain the majority of experimental observations.

In spite of much justifiable criticism, the Flory-Huggins theory can still generate considerable interest because of the limited amount of success which can be claimed for it in relation to phase equilibria studies.

7.3 Phase equilibria

Use can be made of the Flory-Huggins theory to predict the equilibrium behaviour of two liquid phases when both contain amorphous polymer and one or even two solvents.

Consider a two component system consisting of a liquid (1) which is a poor solvent for a polymer (2). Complete miscibility occurs when the Gibbs free energy of mixing is less than the Gibbs free energies of the components, and the solution maintains its homogeneity only as long as ΔG^m remains less than the Gibbs free energy of any two possible co-existing phases.

The situation is represented by curve T_4 in figure 7.1. The miscibility of this type of system is observed to be strongly temperature dependent and as T decreases the solution separates into two phases. Thus at any temperature, say T_1, the Gibbs free energy of any mixture, composition x_2''' in the composition range x_2' to x_2'', is higher than either of the two co-existing phases whose compositions are x_2' and x_2'' and phase separation takes place. The compositions of the two phases x_2' and x_2'' do not correspond to the two minima, but are measured from the points of contact of the double tangent AB with the Gibbs free energy curve. The same is true for other temperatures lying below T_c, and the inflexion points can be joined to bound an area representing the heterogeneous two phase system, where there is limited solubility of component 2 in 1 and vice-versa. This is called a *cloud-point curve.*

As the temperature is increased the limits of this two phase coexistence contract, until eventually they coalesce to produce a homogeneous, one phase,. mixture at T_c, the *critical solution temperature.* This is sometimes referred to as the *critical consolute point.*

In general, we can say that if the free energy-composition curve has a shape which allows a tangent to touch it at two points, phase separation will occur.

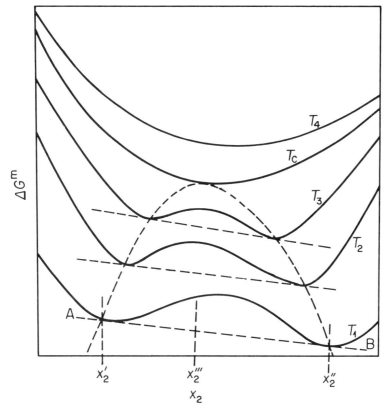

FIGURE 7.1. Schematic diagram of the Gibbs free energy of mixing ΔG^m as a function of the mole fraction x_2 of solute, showing the transition from a system miscible in all proportions (T_4), through the critical temperature T_c to partially miscible systems T_1 to T_3.

The critical solution temperature is an important quantity and can be accurately defined in terms of the chemical potential. It represents the point at which the inflexion points on the curve merge, and so it is the temperature where the first, second, and third derivatives of the Gibbs free energy with respect to mole fraction are zero.

$$\partial(\Delta G^m)/\partial x_2 = \partial^2(\Delta G^m)/\partial x_2^2 = \partial^3(\Delta G^m)/\partial x_2^3 = 0. \qquad (7.4)$$

It is also true that the partial molar Gibbs free energies of each component are equal at this point and it emerges that the conditions for incipient phase separation are

$$\partial\mu_1/\partial\phi_2 = \partial^2\mu_1/\partial\phi_2^2 = \partial^3\mu_1/\partial\phi_2^3 = 0. \qquad (7.5)$$

By remembering that $\Delta G_1 = (\mu_1 - \mu_1^\circ)$, application of these criteria for equilibrium to equation (7.3) leads to the first derivative of that equation

$$(1 - \phi_{2,c})^{-1} - (1 - 1/x_n) - 2\phi_{2,c}\chi_{1,c} = 0, \qquad (7.6)$$

while the second derivative is

$$(1 - \phi_{2,c})^{-2} - 2\chi_{1,c} = 0, \qquad (7.7)$$

where the subscript c denotes critical conditions. The critical composition at which phase separation is first detected is then

$$\phi_{2,c} = 1/(1 + x_n^{1/2}) \approx 1/x_n^{1/2}, \qquad (7.8)$$

and

$$\chi_{1,c} = \tfrac{1}{2} + 1/x_n^{1/2} + 1/2x_n, \qquad (7.9)$$

which indicates that $\chi_{1,c} = 0.5$ at infinitely large chain length.

The interaction parameter χ_1 is a useful measure of the solvent power. Poor solvents have values of χ_1 close to 0.5 while an improvement in solvent power lowers χ_1. Generally, a variation from 0.5 to -1.0 can be observed although for many synthetic polymer solutions the range is 0.6 to 0.3. Measurement of χ_1 can be carried out in a number of ways. If equation (7.3) is expanded in a Taylor series, remembering that the polymer concentration $c_2 = \phi_2 \rho_2$ where ρ_2 is the polymer density, then

$$\Delta G_1 = -RTc_2/x_n\rho_2 + RT(\chi_1 - \tfrac{1}{2})c_2^2/\rho_2^2 - RTc_2^3/\rho_2^3 \qquad (7.10)$$

and for $\chi_1 = 0.5$, ΔG_1 is reduced to the ideal term only.

Inspection of equations (8.4) and (8.10) shows that χ_1 can be related to the second virial coefficient A_2 by $A_2 = (\tfrac{1}{2} - \chi_1)/V_1\rho_2^2$ and calculated from a colligative property or light scattering measurement.

A linear temperature dependence is also predicted for χ_1

$$\chi_1 = a + b/T \qquad (7.11)$$

and equations (7.9) and (7.11) are both relevant to the problems of polymer fractionation.

7.4 Fractionation

The relations derived in this and other chapters normally assume that the polymer sample has a unique molar mass. This situation is rarely achieved in practice and it is useful to know the form of the molar mass distribution in a sample, as this can have a significant bearing on the physical properties. It is also advantageous to be able to prepare sample fractions, whose homogeneity is considerably better than the parent polymer, especially when testing dilute solution theory.

We have seen that the chain length can be related to the solvent power, expressed as χ_1, by equation (7.9) and this is illustrated in figure 7.2. The implication is that if χ_1 can be carefully controlled, conditions could be attained which would allow a given molecular species to precipitate, while leaving larger or smaller molecules in solution. This process is known as *fractionation*.

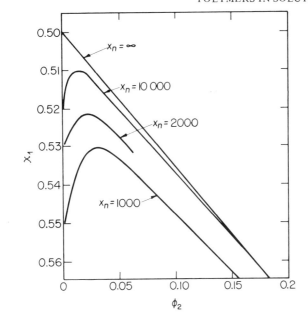

FIGURE 7.2. Variation of χ_1 with volume fraction ϕ_2 of the polymer in solution, showing the effect of chain length x_n.

Experimentally, a polymer sample can be fractionated in a variety of ways and three in common use are: (1) addition of a non-solvent to a polymer solution; (2) lowering the temperature of the solution; and (3) column chromatography.

In the first method the control of χ_1 is effected by adding a non-solvent to the polymer solution. If the addition is slow, χ_1 increases gradually until the critical value for large molecules is reached. This causes precipitation of the longest chains first and these can be separated from the shorter chains which remain in solution. In practice the polymer solution is held at a constant temperature while precipitant is added to the stirred solution. When the solution becomes turbid the mixture is warmed until the precipitate dissolves. The solution is then returned to the original temperature and the precipitate which reforms is allowed to settle and then separated. This ensures that the precipitated fraction is not broadened by local precipitation during addition of the non-solvent. Successive additions of small quantities of non-solvent to the solution allows a series of fractions of steadily decreasing molar mass to be separated.

In the second method, χ_1 is varied by altering the temperature, with similar results. For both techniques, it is useful to dissolve the polymer initially in a poor solvent with a large χ_1 value. This ensures that only small quantities of non-solvent are required to precipitate the polymer in method 1, and that the temperature changes required in method 2 are small.

In column chromatography the polymer is precipitated on the inert support medium at the top of a column which has a temperature gradient imposed along its length. The packing is usually glass beads of 0.1 to 0.3 mm diameter. A solvent + non-solvent mixture is used to elute the sample and fractionation is achieved by using a solvent gradient. This is generated in a mixing system, situated above the column, by constantly increasing the solvent to non-solvent ratio and as the mixture is initially a poor solvent which is gradually enriched by the good solvent the low molar mass fractions are eluted first. Fractions of increasing molar mass are collected from the bottom of the column.

In each of the techniques, the mass and molar mass of the fractions are recorded and a distribution curve for the sample can be constructed from the results. However, as the fractions themselves have a molar mass distribution, extensive overlapping of the fractions will occur as shown schematically in figure 7.3. Consequently a simple histogram constructed from the mass and molar mass of each fraction will not provide a good representation of the distribution and a method must be used to compensate for the overlapping.

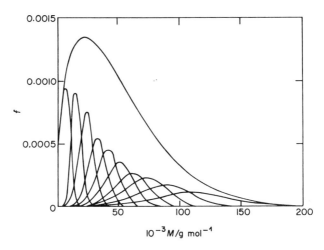

FIGURE 7.3. Schematic representation of the overlapping molar mass distributions of fractions f obtained from the parent sample.

A useful approach was proposed by Schulz who suggested that a cumulative mass fraction be plotted against the molar mass. The cumulative mass fraction $C(M_i)$ can be calculated by adding half the mass fraction w_i of the ith fraction to the total mass fraction of those fractions preceding it, *i.e.*

$$C(M_i) = (w_i/2) + \sum_{j=1}^{i-1} w_j. \qquad (7.12)$$

The values of $C(M_i)$ are plotted against the corresponding M_i and connected by a

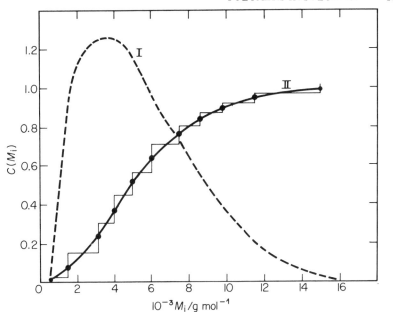

FIGURE 7.4. The differential distribution I, and integral distribution II obtained using equation (7.12).

smooth curve as shown in figure 7.4, to give the integral distribution curve. The differential curve can be obtained by determining the slope of this curve at selected molar masses and plotting this against the appropriate molar mass.

7.5 Flory-Krigbaum theory

To overcome the limitations of the lattice theory resulting from the discontinuous nature of a dilute polymer solution, Flory and Krigbaum discarded the idea of a uniform distribution of chain segments in the liquid. Instead they considered the solution to be composed of areas containing polymer which were separated by the solvent. In these areas the polymer segments were assumed to possess a Gaussian distribution about the centre of mass, but even with this distribution the chain segments still occupy a finite volume from which all other chain segments are excluded. It is within this excluded volume that the long range interactions originate which are discussed more fully in chapter 9.

The thermodynamic functions derived by Flory and Krigbaum to describe these long range effects are given in terms of the excess partial molar quantities as

$$\Delta H_1^E = R T \kappa_1 \phi_2^2 ; \tag{7.13a}$$

$$\Delta S_1^E = R \psi_1 \phi_2^2 ; \tag{7.13b}$$

$$\Delta G_1^E = R T (\kappa_1 - \psi_1) \phi_2^2 . \tag{7.13c}$$

By comparing equation (7.10) with (7.13c) and equating the non-ideal or excess terms we arrive at

$$(\kappa_1 - \psi_1) = (\chi_1 - \tfrac{1}{2}). \tag{7.14}$$

It has already been indicated that the deviations from ideality in a polymer solution can be eliminated by selecting a temperature for which $\Delta H_1^E = T\Delta S_1^E$, or in Flory-Krigbaum terms when $\kappa_1 = \psi_1$. The temperature at which these conditions prevail is called the FLORY or THETA temperature Θ, and is conveniently defined as $\Theta = T\kappa_1/\psi_1$. Substitution in (7.13c) shows that

$$(\kappa_1 - \psi_1) = \psi_1(\Theta/T - 1) = (\chi_1 - \tfrac{1}{2}). \tag{7.15}$$

The theta temperature is a well defined state of the polymer solution at which the excluded volume effects are eliminated and the polymer coil is in an unperturbed condition (see chapter 9). Above the theta temperature expansion of the coil takes place, caused by interactions with the solvent, whereas below Θ the polymer segments attract one another, the excluded volume is negative, and eventual phase separation occurs.

7.6 Location of the theta temperature

The theta temperature of a polymer-solvent system can be measured from phase separation studies. The value of $\chi_{1,c}$ at the critical concentration is related to the chain length of the polymer by equation (7.9), and substitution in (7.15) leads to

$$\psi_1(\Theta/T_c - 1) = 1/x_n^{1/2} + 1/2x_n, \tag{7.16}$$

which on rearrangement gives

$$1/T_c = (1/\Theta)\{1 + (1/\psi_1)(1/x_n^{1/2} + 1/2x_n)\}. \tag{7.17}$$

Remembering that $x_n = (M\bar{v}_2/V_1)$, where M and $\bar{v}_2$ are the molar mass and partial specific volume of the polymer, and V_1 is the molar volume of the solvent, the equation states that the critical temperature is a function of M and the value of T_c at infinite M is the theta temperature for the system.

Precipitation data for several systems have proved the validity of equation (7.17). Linear plots are obtained with a positive slope from which the entropy parameter ψ_1 can be calculated. Typical values are shown in table 7.1, but ψ_1 measured for systems such as polystyrene + cyclohexane have been found to be almost ten times larger than derived from other methods of measurement. This appears to arise from the assumption in the Flory-Huggins theory that χ_1 is concentration independent and improved values of ψ_1 are obtained when this is rectified.

The theta temperature, calculated from equation (7.17) for each system is in good agreement with that measured from the temperature variation of A_2; (see chapter 8). Curves of A_2, measured at various temperatures in the vicinity of Θ, are constructed as a function of temperature for one or more molar masses as shown in figure 7.6. Intersection of the curves with the T-axis occurs when $A_2 = 0$ and $T = \Theta$. The curves for each molar mass of the same polymer should all intersect at $T = \Theta$.

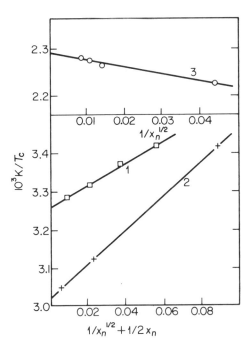

FIGURE 7.5. Chain length x_n dependence of the upper critical consolute temperature T_c for 1, polystyrene in cyclohexane, and 2, polyisobutylene in diisobutyl ketone (data of Schultz and Flory); and the lower critical solution temperature for 3, polyoctene-1 in n-pentane (data of Kinsinger and Ballard.)

TABLE 7.1. Theta temperatures and entropy parameters for some polymer + solvent systems, derived from equation (7.17)

Polymer	Solvent	Θ/K	ψ_1
1. Polystyrene	Cyclohexane	307.2	1.056
2. Polyethylene	Nitrobenzene	503	1.090
3. Polyisobutene	Di-isobutyl Ketone	333.1	0.650
4. Poly(methylmethacrylate)	4-Heptanone	305	0.610
5. Poly(acrylic acid)	Dioxan	302.2	−0.310
6. Polymethacrylonitrile	Butanone	279	−0.630

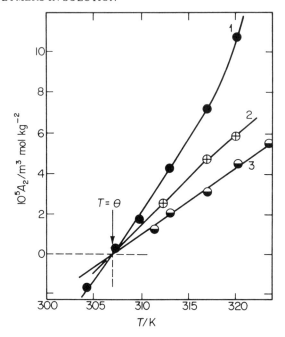

FIGURE 7.6. Location of the theta temperature Θ for poly(α-methyl styrene) in cyclohexane. Values of A_2 are measured for: 1, $M_w = 8.6 \times 10^4$ g mol^{-1}, 2, $M_w = 3.8 \times 10^5$ g mol^{-1}; and 3, $M_w = 1.5 \times 10^6$ g mol^{-1}.

7.7 Lower critical solution temperatures

So far we have been concerned with non-polar solutions of amorphous polymers, whose solubility is increased with rising temperature, because the additional thermal motion helps to decrease attractive forces between like molecules, and encourages energetically less favourable contacts. The phase diagram for such systems, when the solvent is poor, is depicted by area A in figure 7.7, where the critical temperature T_c occurs near the maximum of the cloud-point curve and is often referred to as the *upper critical solution temperature* (UCST).

For non-polar systems ΔS^m is normally positive but weighted heavily by T and so solubility depends mainly on the magnitude of ΔH^m, which is normally endothermic (positive). Consequently as T decreases ΔG^m eventually becomes positive and phase separation takes place.

Values of Θ and ψ_1, in table 7.1, show that for systems 1 to 4 the entropy parameter is positive, as expected, but for poly(acrylic acid) in dioxan and polymethacrylonitrile in butanone, ψ_1 is negative at the theta temperature. As $\psi_1 = \kappa_1$ when $T = \Theta$, the enthalpy is also negative for these systems. This means that systems 5 and 6 exhibit an unusual decrease in solubility as the temperature rises, and the cloud-point curve is now inverted as in area B. The corresponding critical temperature is located at the minimum of the miscibility curve and is known as the *lower critical solution temperature* (LCST).

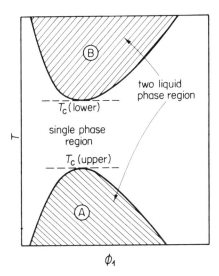

FIGURE 7.7. Schematic diagram of the two types of phase diagram encountered in polymer solutions; A, the two-phase region yielding the upper critical solution temperature; and B, the two-phase region giving the lower critical solution temperature.

In systems 5 and 6 this phenomenon is a result of hydrogen-bond formation between the polymer and solvent, which enhances the solubility. As hydrogen bonds are thermally labile a rise in T reduces the number of bonds and causes eventual phase separation. In solutions, which are stabilized in this way by secondary bonding, the LCST usually appears below the boiling temperature of the solvent but it has been found experimentally that an LCST can be detected in non-polar systems when these are examined at temperatures approaching the critical temperature of the solvent. Polyisobutylene in a series of n-alkanes, polystyrene in methyl acetate and cyclohexane, and cellulose acetate in acetone all exhibit LCSTs.

The LCST is located by heating the solutions, in sealed tubes, up to temperatures approaching the gas-liquid critical point of the solvent. As the temperature rises, the liquid expands much more rapidly than the polymer, which is restrained by the covalent bonding between its segments. At high temperatures, the spaces between the solvent molecules have to be reduced if mixing is to take place. Using the analogy of the lattice, this means that deformation of the lattice must occur to fit both polymer and solvent molecules, and this is substantiated by the negative volumes of mixing observed for such systems. The negative enthalpy, associated with the LCST, arises from the effective condensation of the solvent in the polymer. This contributes to a negative Gibbs free energy and improves the solubility as the temperature rises, but now the overall solubility is dominated by the negative entropy and large T.

Two effects contribute to ΔS^m, (i) combinatorial mixing and (ii) molecular ordering arising from the contraction of the expanded solvent to fit the lattice. This non-combinatorial term is negative and unfavourable. As the temperature rises, it becomes increasingly difficult for the solvent to contract on to the polymer lattice, and the non-combinatorial term soon begins to predominate. When this occurs ΔG^m becomes positive and phase separation takes place.

The Flory-Huggins theory cannot predict the LCST, because it does not include a non-combinatorial entropy term. Equation (7.11) indicates that the dependence of χ_1 on T is linear, and predicts an ever decreasing χ_1 as T increases.

A more elaborate treatment of polymer solutions by Prigogine, considers differences in the thermal expansion of the components of the solution, and has met with some success in describing the LCST. The new form for χ_1 is now

$$\chi_1 = A_0 r_1 / RT + B_0 T / r_1 R, \qquad (7.18)$$

where r_1 is the number of segments in the solvent and A_0 and B_0 are molecular parameters characteristic of a series of systems. This now predicts that as T increases χ_1 passes through a minimum and eventually regains its critical value at the LCST.

For a polymer of infinite molar mass, the LCST will correspond to a "lower" theta temperature where the macromolecule regains its unperturbed dimensions. In the pseudo-ideal state of the LCST, however, the molecular contraction is encouraged by the negative and unfavourable entropy of dilution, not the enthalpy, as was the case at the UCST.

7.8 Solubility and the cohesive energy density

Solvent-polymer compatibility problems are repeatedly encountered in industry. For example, in situations requiring the selection of elastomers for use as hose-pipes or gaskets, the correct choice of elastomer is of prime importance, as contact with highly compatible fluids may cause serious swelling and impair the operation of the system. The wrong selection can have far reaching consequences; the initial choice of an elastomer for the seals in the landing gear of the DC-8 aircraft resulted in serious jamming because the seals became swollen when in contact with the hydraulic fluid. This almost led to grounding of the plane but replacement with an incompatible elastomer made from ethylene-propylene copolymer rectified the fault.

To avoid such problems a technologist may wish to have at his disposal a rough guide to aid the selection of solvents for a polymer or to assess the extent of polymer-liquid interaction other than those already described. Here use can be made of a semi-empirical approach suggested by Hildebrand and based on the premise that "like dissolves like". The treatment involves relating the enthalpy of mixing to the cohesive energy density (E/V) and defines a solubility parameter $\delta = (E/V)^{1/2}$, where E is the molar energy of vaporization and V is the molar volume of the component. The proposed relation, for non-polar systems,

$$\Delta H_1 = (\delta_1 - \delta_2)^2 V_1 \phi_2^2, \qquad (7.19)$$

shows that ΔH_1 is small for mixtures with similar solubility parameters and this indicates compatibility.

Values of the solubility parameter for simple liquids can be readily calculated from the enthalpy of vaporization. The same method cannot be used for a polymer and one must resort to comparative techniques. Usually δ for a polymer is established by finding the solvent which will produce maximum swelling of a network or the largest value of the limiting viscosity number, as both indicate maximum compatibility. The polymer is then assigned a similar value of δ. Alternatively, Small has tabulated a series of group molar attraction constants from which a good estimate of δ for most polymers can be made.

Attempts to correlate δ with χ_1 from the Flory-Huggins equation have met with limited success because of the unjustifiable assumptions made in the derivation. It is now believed, however, that χ_1 is not an enthalpy parameter but a free energy parameter, as shown in equation (7.4), and a relation of the form

$$\chi_1 = 1/z + (V_1/RT)(\delta_1 - \delta_2)^2, \qquad (7.20)$$

has improved the correlation. Here $1/z = \chi_s$ is supposed to compensate for the lack of a non-combinatorial entropy contribution in the Flory-Huggins treatment.

Unfortunately, solubility is not a simple process and secondary bonding may play an important role in determining component interactions. This is illustrated in figure 7.8 for the elastomer SBR. The diagram represents a contour map of elastomer-liquid swelling where the hydrogen-bonding index γ of the liquid is also considered. It can be seen that there is 100 per cent swelling of the elastomer in liquids with δ in the range 7.7 to 10.7 for which $\gamma = 0$, but the swelling is restricted to 25 per cent in the same δ range when $\gamma > 19.0$. Thus the maximum information is derived only when both parameters are considered.

The hydrogen-bonding index γ is derived on the basis of work by W. Gordy. He measured the infrared spectrum of a solution of deuterated methanol (CH_3OD) in benzene (1 mol dm^{-3}), and used the absorption peak at 2681 cm^{-1} as a standard reference. When benzene is replaced by other liquids the peak is displaced and the extent of the shift is a measure of γ, e.g. for acetone the peak is at 2584 cm^{-1} and so $\gamma = (\Delta \bar{\nu}/10 \text{ cm}^{-1}) = 9.7$. Typical values are shown in table 7.2 and δ for some polymers in table 7.3.

Other more detailed approaches have been suggested and Hansen introduced a three-dimensional δ composed of contributions from van der Waals dispersion forces, dipole-dipole interactions, and hydrogen bonding. However, one must consider whether such elaboration of an essentially semi-empirical concept is justified, especially when the evidence suggests that dipolar interactions can normally be dispensed with.

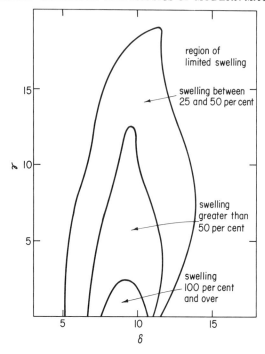

FIGURE 7.8. "Swelling contour map" for SBR. (From data by Beerbower, Kaye, and Pattison, (1967).

TABLE 7.2. Solubility parameters δ and hydrogen bonding indices γ for some common solvents; $cal_{th} = 4.184$ J

Solvent	γ	$\delta/(cal_{th}cm^{-3})^{1/2}$	Solvent	γ	$\delta/(cal_{th}cm^{-}$
Acetone	9.7	10.0	Dioxan	9.7	9.9.
Aniline	18.1	11.8	Ethylenedichloride	1.5	9.8
Benzene	0.0	9.2	Ethyl ether	13.0	7.4
Benzaldehyde	8.4	9.4	Ethylene glycol	20.6	14.2
Butyl acetate	8.8	8.5	Hexane	0.0	7.3
Butylamine	16.8	9.1	Methyl alcohol	18.7	14.5
Carbon disulphide	0.0	10.0	Methylene dichloride	1.5	9.7
Carbon tetrachloride	0.0	8.6	Nitrobenzene	2.8	10.0
Chloroform	1.5	9.3	Nitromethane	2.5	12.7
Cyclohexane	0.0	8.2	Pyridine	18.1	10.7
Cyclohexanone	11.7	9.3	Tetrachloroethylene	1.5	9.3
Cyclohexanol	18.7	11.4	Glycerol	22.0	16.5
Dichloroethylene	1.5	9.3	Water	39.0	23.4

TABLE 7.3. Solubility parameters for common polymers; cal_{th} = 4.184 J

Polymer	$\delta/(cal_{th}\ cm^{-3})^{1/2}$	Polymer	$\delta/(cal_{th}\ cm^{-3})^{1/2}$
Teflon	6.2	Polystyrene	9.15
Poly(isobutylene)	7.7	Poly(methyl metha-crylate)	9.3
Poly(isoprene)	8.15	Poly(vinyl acetate)	9.4
Polyethylene	8.1	Poly(ethylene terephthalate)	10.7
Polypropylene	8.1	Nylon 6,6	13.6
Poly(vinyl chloride)	8.9	Polyacrylonitrile	15.4

General Reading

F. W. Billmeyer, *Textbook of Polymer Science*, Chapter 2. John Wiley and Sons (1962).

M. J. R. Cantow, *Polymer Fractionation*. Academic Press (1967).

P. J. Flory, *Principles of Polymer Chemistry*, Chapters 12 and 13. Cornell University Press, Ithaca, N.Y. (1953).

J. H. Hildebrand and R. L. Scott, *Regular Solutions*. Prentice-Hall (1962).

M. L. Miller, *The Structure of Polymers*, Chapter 4. Reinhold Publishing Corp. (1968).

H. Morawetz, *Macromolecules in solution*. Interscience Publishers Inc. (1965).

D. A. Smith, *Addition Polymers*, Chapter 6. Butterworths (1968).

A. V. Tobolsky and H. Mark, *Polymer Science and Materials*, Chapter 4. Wiley-Interscience (1971).

H. Tompa, *Polymer Solutions*. Butterworths (1956).

References

1. A. Beerbower, L. A. Kaye, and D. A. Pattison, *Chemical Engineering*, 118, (1967).
2. J. B. Kinsinger and L. E. Ballard, *Polymer Letters*, 2, 879 (1964).
3. A. R. Schultz and P. J. Flory, *J. Amer. Chem. Soc.*, 74, 4760 (1952).

Chapter 8

Polymer Characterization—Molar Masses

8.1 Introduction

Many of the distinctive properties of polymers are a consequence of the long chain lengths, which are reflected in the large molar masses of these substances. While such large molar masses are now taken for granted, it was difficult in 1920 to believe and accept that these values were real and not just caused by the aggregation of much smaller molecules. Values of the order of 10^6 g mol^{-1} are now accepted without question, but the accuracy of the measurements is much lower than for simple molecules. This is not surprising, especially when polymer samples exhibit polydispersity, and the molar mass is, at best, an average dependent on the particular method of measurement used. Estimation of the molar mass of a polymer is of considerable importance, as the chain length can be a controlling factor in determining solubility, elasticity, fibre forming capacity, tear strength, and impact strength in many polymers.

The methods used to determine the molar mass M are either relative or absolute. Relative methods require calibration with samples of known M and include viscosity and vapour pressure osmometry. The absolute methods are often classified by the type of average they yield, *i.e.* colligative techniques yield number averages, light scattering and the ultracentrifuge yield higher averages, *e.g.* weight and z-average.

8.2 Molar masses, molecular weights, and SI units

The dimensionless quantity "the relative molecular mass" (molecular weight) defined as the average mass of the molecule divided by 1/12th the mass of an atom of the nuclide C^{12}, is often used in polymer chemistry, and called the molecular weight. In this book the quantity molar mass is used and appropriate SI units are given.

8.3 Number average molar mass M_n

Determination of the number average molar mass M_n involves counting the total number of molecules, regardless of their shape or size, present in a unit mass of the polymer. The methods are conveniently grouped into three categories: end-group assay, thermodynamic, and transport methods.

8.4 End-group assay

The technique is of limited value and can only be used when the polymer has an end group amenable to analysis. It can be used to follow the progress of linear condensation reactions when an end group, such as a carboxyl, is present which can be titrated. It is used to detect amino end groups in nylons dissolved in m-cresol, by titration with methanolic perchloric acid solution, and can be applied to vinyl polymers if an initiator fragment, perhaps containing halogen, is attached to the end of the chain.

The sensitivity of the method decreases rapidly as the chain length increases and the number of end groups drops. A practical upper limit might reach an M_n of about 15000 g mol^{-1}.

8.5 Thermodynamic methods

Because chemical methods are rather limited, it is necessary to turn to physical measurements to obtain the scope required for the wide variety of polymers not suitable for end-group analysis. Several techniques are available which are based on the colligative properties of dilute solutions. A colligative property is defined as one which is a function only of the number, and not of the nature, of the solute molecules contained in a unit volume of solution. Clearly then, colligative properties such as osmotic pressure, elevation of boiling temperature, depression of freezing temperature, and lowering of vapour pressure, should provide a convenient means of establishing M_n. Each technique requires that an equilibrium is established, between the solvent in solution and the pure solvent, which is either in another phase or separated from the solution by a semi-permeable barrier. Consequently these properties are determined by the solvent activity, and Raoult's law can be used. As polymer solutions are non-ideal they will only obey Raoult's law if the solutions are very dilute. Under these conditions, the molar mass M_2 of the dissolved polymer (component 2) can be calculated from the solvent activity a_1 using

$$a_1 = x_1 = (w_1/M_1)/ \{ (w_1/M_1) + (w_2/M_2) \} = n_1/(n_1 + n_2) \qquad (8.1)$$

where n_1 and n_2 are the amounts and w_1 and w_2 the masses of solvent and polymer respectively.

When the solution is sufficiently dilute

$$1 - a_1 \approx \{ (w_2/M_2)/(w_1/M_1) \}, \qquad (8.2)$$

and any of the colligative methods can be used to determine M_2 using this approximation.

8.6 Ebullioscopy and cryoscopy

In principle these two methods can be treated together. The relevant expressions are derived from the Clausius-Clapeyron equation and the resulting expression is

$$M_n = (RT^2 V_1/\Delta H)(c/\Delta T)_{c \to 0}. \tag{8.3}$$

For ebulliometry, T, ΔH, and ΔT are the boiling temperature of the solvent, the enthalpy of vaporization of the solvent, and the elevation of the boiling temperature respectively, while for cryoscopy they represent the freezing temperature of the solvent, the enthalpy of fusion of the solvent, and the depression of the freezing temperature. The equation represents the limiting case at infinite dilution and it is necessary to extrapolate $(\Delta T/c)$ for a series of solutions to $c = 0$ in order to calculate M_n.

The measurements are limited by the sensitivity of the thermometer used to obtain ΔT. At present this can rarely detect a ΔT of less than 1×10^{-3} K with any precision, and the limit of accurate measurement of M_n is in the region of 25000 to 30000 g mol^{-1}.

8.7 Osmotic pressure

Measurement of the osmotic pressure π of a polymer solution can be carried out in the type of cell represented schematically in figure 8.1. The polymer solution

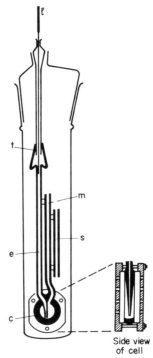

FIGURE 8.1. Pinner Stabin osmometer, consisting of glass cell c, measuring capillary m, reference capillary s, filling tube e, levelling rod l, and mercury cup t.

is separated from the pure solvent by a membrane, permeable only to solvent molecules. Initially, the chemical potential μ_1, of the solvent in the solution, is lower than that of the pure solvent μ_1°, and solvent molecules tend to pass through the membrane into the solution in order to attain equilibrium. This causes a build up of pressure in the solution compartment until, at equilibrium, the pressure exactly counteracts the tendency for further solvent flow. This pressure is the osmotic pressure.

For dilute polymer solutions

$$(\mu_1 - \mu_1^\circ) = RT \ln a_1 = -V_1 \pi. \tag{8.4}$$

In the limit of infinite dilution, this can be related to the ratio of the amounts of each component,

$$\pi V_1 = RT \ln \bar{x}_2 \approx RT \ln (n_2/n_1) \tag{8.5}$$

and since $n_1 V_1$ is simply the volume of the solvent and $\ln (n_2/n_1) \approx (n_2/n_1)$ then

$$\pi V/n_1 = RT(n_2/n_1),$$

or

$$\pi V = n_2 RT. \tag{8.6}$$

The equation can be expressed in terms of the concentration when we substitute $c = \Sigma(N_i M_i)/N_A V$ and $n_2 = \Sigma N_i/N_A$ giving

$$(\pi/c)\Sigma(N_i M_i)/N_A = RT\Sigma N_i/N_A, \tag{8.7}$$

or

$$(\pi/c)_{c \to 0} = RT/M_n, \tag{8.8}$$

where N_A is the Avogadro constant. Equation (8.8) is again a limiting form, valid at infinite dilution.

Only under special conditions, when the polymer is dissolved in a theta-solvent, will (π/c) be independent of concentration. Experimentally, a series of concentrations is studied and the results treated according to one or other of the following virial expansions. McMillan and Meyer suggested,

$$\pi/c = RT/M_n + Bc + Cc^2 + \ldots, \tag{8.9}$$

while alternative forms are also used:

$$\pi/c = RT(1/M_n + A_2 c + A_3 c^2 + \ldots); \tag{8.10}$$

and

$$\pi/c = (\pi/c)_0 (1 + \Gamma_2 c + \Gamma_3 c^2 + \ldots). \tag{8.11}$$

The coefficients B, A_2, Γ_2 and C, A_3, Γ_3, are the second and third virial coefficients. When solutions are sufficiently dilute a plot of (π/c) against c is linear and the third virial coefficients (C, A_3, Γ_3) can be neglected. The various forms of the second virial coefficient are interrelated by

$$B = RTA_2 = RT\Gamma_2/M_n. \tag{8.12}$$

Although not normally detected, the third virial coefficient occasionally contributes to the non-ideal behaviour in dilute solutions and a curved plot is obtained. (Figure 8.2a.) This increases the uncertainty of the extrapolation, but can be overcome by recasting equation (8.11) and introducing a polymer-solvent interaction parameter g

$$\pi/c = (RT/M_n)(1 + \Gamma_2 c + g\Gamma_2^2 c^2). \tag{8.13}$$

It has been found that $g = 0.25$ in good solvents so that equation (8.13) becomes

$$\pi/c = (RT/M_n)(1 + \tfrac{1}{2}\Gamma_2 c)^2. \tag{8.14}$$

A plot of $(\pi/c)^{1/2}$ against c is now linear and this extrapolation is illustrated in figure 8.2b.

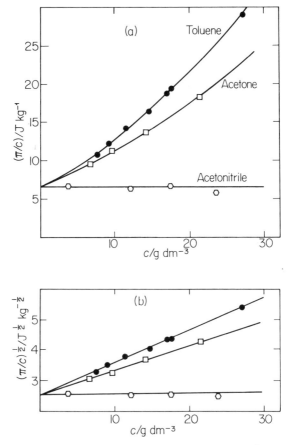

FIGURE 8.2. (a) Plot of (π/c) against c for a sample of poly(methyl methacrylate) $M_n = 382\,000$ g mol^{-1} in three solvents. (b) Plot $(\pi/c)^{1/2}$ against c for the same data as in (a). (From data by Fox et al., 1962.)

This example (figure 8.2) also shows the differing solubility of poly(methyl methacrylate) in the three solvents. In a good solvent, toluene, the slope or A_2 is large, but as the solvent becomes poorer (acetone) A_2 decreases, until it is zero in the theta-solvent acetonitrile. Thus A_2 provides a useful measure of the thermodynamic quality of the solvent and measures the deviation from ideality of the polymer solution.

The value of M_n is calculated from the intercept $(\pi/c)_0 = 6.4$ J kg^{-1} using equation (8.8).

$$M_n = RT/(\pi/c)_0 = 8.314 \text{ J K}^{-1} \text{ mol}^{-1} \times 303 \text{ K}/6.4 \text{ J kg}^{-1} = 393.62 \text{ kg mol}^{-1}.$$

The corresponding values of the second virial coefficient are obtained from the slopes of the plots (table 8.1).

TABLE 8.1. Values of B, A_2 and Γ_2 for poly(methyl methacrylate) dissolved in three solvents

Solvent	B/J m^3 kg^{-2}	A_2/m^3 mol kg^{-2}	Γ_2/m^3 kg^{-1}
Toluene	0.525	2.08×10^{-4}	8.2×10^{-2}
Acetone	0.410	1.63×10^{-4}	6.4×10^{-2}
Acetonitrile	0	0	0

PRACTICAL OSMOMETRY

The static method of determining the osmotic pressure of a polymer solution, using volumes of 3 to 20 cm^3 of solution, is a relatively slow process which requires about 24 h to equilibrate at each concentration. Several designs, suitable for this type of measurement, are typified by the Pinner Stabin instrument shown schematically in figure 8.1. The osmometer is assembled, under a layer of solvent by clamping two membranes (kept continually moist with solvent) on either side of the glass cell c. These are retained in position by two metal plates perforated and grooved to allow contact between the membrane and solvent which is in the outer container.

The preparation of the membranes is very important and must be carefully carried out. They are normally made of cellulose or a cellulose derivative and should be slowly conditioned from the storage liquid to the solvent in use. This is done by transferring the membrane to mixtures progressively richer in the solvent, allowing them time to equilibrate with the mixture, then transferring again until pure solvent is reached. Equilibration in each mixture usually takes a few hours.

When assembled, the osmometer is placed in a jacket containing enough solvent to cover the bottom part of the reference capilary s. Solvent is then withdrawn from the cell c and a solution of polymer added by means of a syringe. Care is taken during the filling stage to avoid trapping bubbles in the cell. The level of the solution is then adjusted to a few centimetres above the level of solvent in s by means of a levelling rod l. Mercury is added to the cup t, to ensure a leak free system, and the osmometer is left undisturbed in a thermostat bath controlled to ± 0.01 K to reach equilibrium.

The osmotic pressure can be calculated from the difference in heights h between the solvent and solution in s and m respectively and π is measured from $\pi = h\rho g$, for each concentration where ρ is the density of the solution and g the acceleration of free fall. Results are plotted as (π/c) against c as described and M_n is calculated from the intercept. The method suffers from the disadvantage that it is slow and consequently diffusion of low molar mass material could be large enough to introduce serious error in the measurement.

Two or three high-speed automatic membrane-osmometers have now been designed to reduce these drawbacks and are commercially available. The Mechrolab osmometer, shown schematically in figure 8.3, consists of a solution + solvent cell of volume approximately 1 cm^3, with the solvent side connected to a reservoir attached to a servo-driven elevator. When solution is added to the top-half of the cell, solvent in the lower-half tends to flow into the upper section to equalize the chemical potentials. The flow is detected optically by the movement of a bubble in a capillary below the cell. The movement activates the servo-motor, which alters the hydrostatic head thereby counteracting the flow. The movement of the solvent reservoir is then a measure of the osmotic

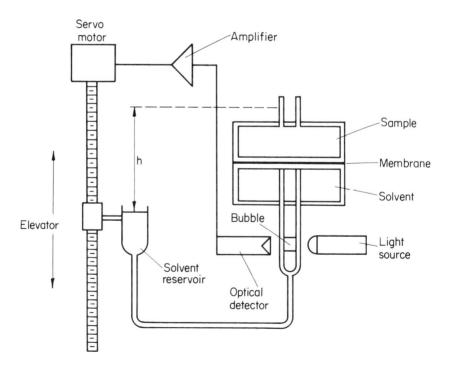

FIGURE 8.3. Schematic diagram of the Mechrolab rapid membrane osmometer.

pressure of the solution. Equilibration is rapid (5 to 30 min) and permeation is readily detected, if present, by following the change of head as a function of time on a recorder. There is no actual flow of solvent in the Mechrolab instrument.

A slightly different principle, which allows solvent flow to take place, forms the basis of the Melab and Knauer models. The Melab osmometer has a stainless steel cell (volume 0.5 cm^3), with solution and solvent compartments separated by the membrane. One wall of the cell is a flexible stainless steel diaphragm connected through a strain gauge to a recorder. As solvent diffuses through the membrane, the increase in volume causes the diaphragm to move. The motion is detected by the gauge and translated into a pressure. The design has the advantage that both solvent and solution compartments are easily rinsed out and the cell does not have to be dismantled if contamination by permeation of low molar mass solute occurs.

All osmotic pressure measurements are extremely sensitive to temperature and must be carried out under rigorously controlled temperature conditions. This is allowed for in each instrument and in addition, measurements can be made over a range of temperatures (278 to 373 K). Solvents should be chosen which are chemically stable and have a low to medium vapour pressure at the temperature of operation, as this avoids problems of bubble formation in the measuring chamber.

8.8 Transport methods – vapour pressure osmometer

In conventional osmometry, the membrane permeability imposes a lower limit of about M_n = 15000 g mol^{-1}. A technique, based on the lowering of the vapour pressure, called vapour pressure osmometry is a useful method of measuring values of M_n from 50 to 20000 g mol^{-1}. It is a relative method and is calibrated using such low molar mass standards as benzil, methyl stearate, or glucose penta-acetate.

The apparatus consists of a thermostatted chamber, saturated with solvent vapour at the temperature of measurement, and containing two differential matched thermistors which are capable of detecting temperature differences as low as 10^{-4} K. Two syringes, one for solvent and one for solution, are used to apply a drop of solution to one thermistor, and a drop of solvent to the other. As there is a difference in vapour pressure between the solution and the solvent drops, solvent from the vapour phase will condense on the solution drop causing its temperature to rise. Because of the large excess of solvent present, evaporation, and hence cooling of the solvent drop, is negligible. When equilibrium is attained, the temperature difference between the two drops ΔT is a measure of the extent of the vapour pressure lowering by the solute. The thermistors form part of a Wheatstone bridge, and ΔT is recorded as a difference in resistance ΔR. The molar mass can then be calculated from

$$\Delta R/K^*c = (1/M_n)(1 + \tfrac{1}{2}\Gamma_2 c)^2, \qquad (8.15)$$

where K^* is the calibration constant. As with other methods M_n is obtained by extrapolating the data to $c = 0$. The calibration constant is estimated by measuring ΔR for solutions of known concentration prepared from standard compounds of known molar mass M_k then

$$K^* = M_k(\Delta R/c)_{c \to 0}. \tag{8.16}$$

In some instances an additional correction for the dilution of the drop of solution may be necessary.

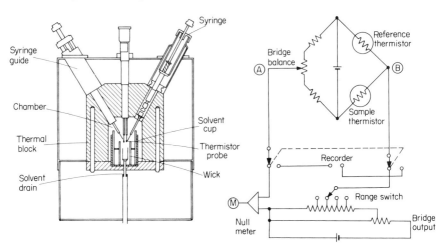

FIGURE 8.4. Sample chamber and circuit diagram for a typical vapour pressure osmometer.

8.9 Light scattering

Light scattering is one of the most popular methods for determining the weight average molar mass M_w. The phenomenon of light scattering by small particles is familiar to us all; the blue colour of the sky or the varied colours of a sunset, the poor penetration of car headlights in a fog is caused by water droplets scattering the light, and the obvious presence of dust in a sunbeam or the Tyndall effect in an irradiated colloidal solution are further examples of this effect.

The fundamentals of light scattering were expounded by Lord Rayleigh in 1871 during his studies on gases, where the particle is small compared with the wavelength of the incident radiation. Light is an electromagnetic wave, produced by the interaction of a magnetic and electric field, both oscillating at right angles to one another in the direction of propagation. When a beam of light strikes the atoms or molecules of the medium, the electrons are perturbed or displaced and oscillate about their equilibrium positions with the same frequency as the exciting beam. This induces transient dipoles in the atoms or molecules, which act as secondary scattering centres by re-emitting the absorbed energy in all directions, *i.e.* scattering takes place.

For gases, Rayleigh showed that the reduced intensity of the scattered light R_θ at any angle θ to the incident beam, of wavelength λ could be related to the molar mass of the gas M, its concentration c, and the refractive index increment $(d\bar{n}/dc)$ by

$$R_\theta = (2\pi^2/N_A\lambda^4)(d\bar{n}/dc)^2(1 + \cos^2\theta)Mc. \tag{8.17}$$

The quantity R_θ is often referred to as the Rayleigh ratio and is equal to $(i_\theta r^2/I_0)$ where I_0 is the intensity of the incident beam, i_θ is the quantity of light scattered per unit volume by one centre at an angle θ to the incident beam, and r is the distance of the centre from the observer. This is valid for a gas, where all the particles are considered to be independent scattering centres and the addition of more centres, which increases $\bar{n}$, increases the scattering. The situation changes when dealing with a liquid as $(d\bar{n}/dc)$ remains unaffected by the addition of molecules and can be expected to be zero. This conceptual difficulty was overcome in the fluctuation theories of Smoluchowski and Einstein; they postulated that optical discontinuities exist in the liquid arising from the creation and destruction of holes during Brownian motion. Scattering emanates from these centres, created by local density fluctuations, which produce changes in $(d\bar{n}/dc)$ in any volume element.

When a solute is dissolved in a liquid, scattering from a volume element again arises from liquid inhomogeneities, but now an additional contribution from fluctuations in the solute concentration is present and for polymer solutions the problem is to isolate and measure these additional effects. This was achieved by Debye in 1944, who showed that for a solute whose molecules are small compared with the wavelength of the light used, the reduced angular scattering intensity of the solute is

$$R_\theta = R_\theta(\text{solution}) - R_\theta(\text{solvent}), \tag{8.18}$$

and that this is related to the change in Gibbs free energy with concentration of the solute. As ΔG is related to the osmotic pressure π, we have

$$R_\theta = (2\pi^2\bar{n}_0^2/\lambda^4)(1 + \cos^2\theta)(d\bar{n}/dc)^2(NM/N_A)\{RT/(d\pi/dc)_T\}. \tag{8.19}$$

Here $\bar{n}_0$ and $\bar{n}$ are the refractive indices of solvent and solution respectively, and N is the number of polymer molecules. Differentiation of the virial expansion for π with respect to c, followed by substitution in equation (8.19) and rearrangement leads to

$$K'(1 + \cos^2\theta)c/R_\theta = 1/M_w + 2A_2c, \tag{8.20}$$

where

$$K' = \{2\pi^2\bar{n}_0^2(d\bar{n}/dc)^2/\lambda^4 N_A)\}. \tag{8.21}$$

Alternatively, the scattering can be expressed as a turbidity τ where

$$\tau = (16\pi/3)R_\theta, \tag{8.22}$$

and the equation becomes

$$Hc/\tau = 1/M_w + 2A_2c + \ldots \tag{8.23}$$

The new constant is $H = \{(16\pi/3)K'(1 + \cos^2\theta)\}$. Both equations are valid for molecules smaller than $(\lambda'/20)$ when the angular scattering is symmetrical. Here λ' is the wavelength of light in solution $\lambda' = (\lambda/\bar{n}_0)$.

For small particles, M_w can be calculated from either equation (8.20) or (8.23). The important experimental point to remember is that dust will also scatter light and contribute to the scattering intensity. Great care must be taken to ensure that solutions are clean and free of extraneous matter. Solutions of the polymer are prepared in a concentration series and clarified either by centrifugation for a few hours at about $25000\,g$, or filtered through a grade 5 sinter glass filter. Alternatively, a millipore filter, porosity 0.45 μm can be used.

A number of instruments are available commercially; only one is described here and the schematic diagram 8.5 provides the main features of the model.

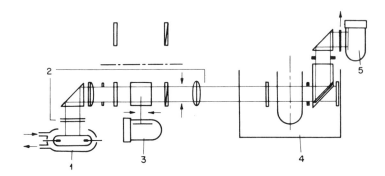

FIGURE 8.5. Schematic representation of the optics of a SOFICA light scattering instrument: 1, the light source, a water cooled mercury vapour lamp; 2, path of incident beam through a system of filters, polarizers, and a variable slit; 3, reference photomultiplier; 4, thermostat; and 5, photomultiplier.

Light is obtained from a water-cooled mercury vapour lamp and one of three wavelengths 365, 436, or 546 nm can be selected by means of an appropriate filter. As the scattering intensity is a function of λ^{-4}, use of a lower wavelength enhances the scattering, but the choice is left to the operator. The light beam, which can be polarized, or left unpolarized, is collimated before passing through the cell. The measuring cell is immersed in a vat of liquid, usually benzene or xylene which can be thermostatted at temperatures between 273 and 400 K. Scattering is detected by a photomultiplier, capable of revolving round the cell and the intensity is recorded on a galvanometer. The 90° scattering (R_{90}) is plotted as $(K''c/R_{90})$ against c and linear extrapolation of the results leads to M_w as the intercept at $c = 0$.

Typical results are shown in table 8.2 for a polystyrene sample dissolved in benzene. The relevant constants are $(d\bar{n}/dc) = 0.112 \times 10^{-3}\,\mathrm{m^3\ kg^{-1}}$, $K' = 2.5888 \times 10^{-5}\,\mathrm{m^2\ mol\ kg^{-2}}$, the intercept $(K'c/R_{90})_{\bar{c}\,=\,0}$ $6.9 \times 10^{-3}\,\mathrm{mol\ kg^{-1}}$ and $M_w = 148\ \mathrm{kg\ mol^{-1}}$.

TABLE 8.2. Polystyrene in benzene at 298 K

$c/\text{g dm}^{-3}$	$10^3 R_{90}/\text{m}^{-1}$	$(K'c/R_{90})/\text{mol g}^{-1}$
1.760	5.31	8.56
3.708	8.43	11.36
6.244	11.24	14.35
7.736	12.43	16.07
10.230	13.80	19.15

SCATTERING FROM LARGE PARTICLES

When polymer dimensions are greater than $\lambda'/20$, intraparticle interference causes the scattered light from two or more centres to arrive considerably out of phase at the observation point, and the scattering envelope becomes dependent on the molecular shape. This attenuation, produced by destructive interference, is zero in the direction of the incident beam, but increases as θ increases because the path length difference $\Delta\lambda_f$ in the forward direction is less than $\Delta\lambda_b$ in the backward, (see figure 8.6). This difference can be measured from the dissymmetry coefficient Z

$$Z = R_\theta/R_{\pi - \theta} \qquad (8.24)$$

which is unity for small particles, but greater than unity for large particles. The scattering envelope reflects the scattering attenuation and is compared with that for small particles in figure 8.7. The angular attenuation of scattering is measured by the particle scattering factor $P(\theta)$ which is simply the ratio of the scattering intensity to the intensity in the absence of interference, measured at the same angle θ.

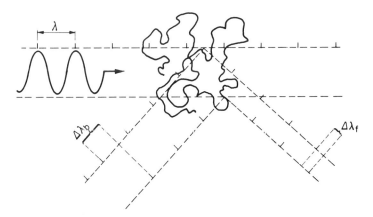

FIGURE 8.6. Destructive interference of light scattered by large particles. Waves arriving at the forward observation point are $\Delta\lambda_f$ out of phase and those arriving at the backward point are $\Delta\lambda_b$ out of phase. For large polymer molecules $\Delta\lambda_f < \Delta\lambda_b$ as shown also in figure 8.7.

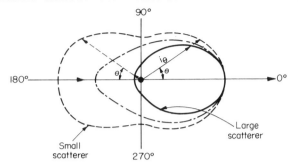

FIGURE 8.7. Intensity distribution of light scattered at various angles. The symmetrical envelope is obtained for small isotropic scatterers in dilute solution, the two assymmetric envelopes are for much larger scattering particles. The solid line represents the scattering from spheres whose diameters are approximately one half of the wavelength of the incident light.

Gunier showed that a characteristic shape-independent geometric function, called the radius of gyration $\langle \tilde{S}^2 \rangle^{1/2}$ can be measured from large particle scattering. It is defined as an average distance from the centre of gravity of a polymer coil to the chain end.

The function $P(\theta)$ is size dependent and can be related to the polymer coil size by

$$P(\theta) = (2/u^2)\{e^{-u} - (1-u)\}, \tag{8.25}$$

where $u = \{(4\pi/\lambda)\sin(\theta/2)\}^2 \langle \tilde{S}^2 \rangle$, for monodisperse randomly coiling polymers. In the limit of small θ the expansion

$$P(\theta)^{-1} = 1 + u/3 - \dots \tag{8.26}$$

can be used, and the coil size can be estimated from $P(\theta)$ without assuming a particular model. Specific shapes can be related to $P(\theta)$ if desired, as shown in figure 8.8a and b.

Two methods can be used to calculate M_w and the particle size for large molecules.

(i) *Dissymmetry method.* If $\mathcal{Z} = \{P(\theta)/P(\pi - \theta)\}$ is not too large, one need only measure the scattering intensity at $90°$ and two angles symmetrical about $90°$, usually $45°$ and $135°$. As $\mathcal{Z}$ is normally concentration dependent, the value at $c = 0$ is obtained by plotting $(\mathcal{Z} - 1)^{-1}$ against c. From published tables $\mathcal{Z}_{c=0}$ can be related to $P(90)$, and M_w is calculated from the $90°$ scattering then corrected by multiplication with $P(90)$. Also available in table form is the ratio $(\langle \tilde{r}^2 \rangle^{1/2}/\lambda')$ presented as a function of $\mathcal{Z}$, where $\langle \tilde{r}^2 \rangle^{1/2}$ is the root mean square distance between the ends of the polymer coil. The corresponding functions for a rod and a sphere have different forms (figure 8.8b). Polymer dimensions can be calculated in this way if an assumption is made about the best model. A much more satisfying treatment of the data uses the double extrapolation method proposed by Zimm, which leads to the shape independent parameter $<\bar{S}^2>^{1/2}$.

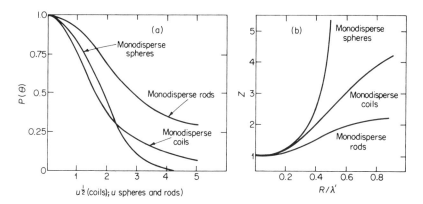

FIGURE 8.8. (a) $P(\theta)$ as a function of u for three model shapes; coils, spheres, and rods. (b) The dissymmetry Z as a function of R/λ', where R is the characteristic linear dimension of either a rod, a sphere, or a coil.

(ii) *Zimm plots.* This is based on the knowledge that, as the scattering at zero angle is independent of size, $P(\theta)$ is unity when $\theta = 0$. Experimentally this is difficult to measure, and an extrapolation procedure has been devised which makes use of a modified form of equation (8.20) for large particles,

$$Kc/R_\theta = 1/M_w P(\theta) + 2A_2 c + \dots \quad (8.27)$$

Substituting for $P(\theta)$ leads to

$$Kc/R_\theta = 1/M_w + (1/M_w)\{(16\pi^2/3\lambda'^2)\sin^2(\theta/2)<\bar{S}>_z^2\} + 2A_2 c + \dots \quad (8.28)$$

If the scattering intensity for each concentration in a dilution series is measured over an angular range $35°$ to $145°$, the data can be plotted as (Kc/R_θ) against $\{\sin^2(\theta/2) + k'c\}$, where k' is an arbitrary constant chosen to provide a convenient spread of the data in the grid-like graph which is obtained. A double extrapolation is then carried out, as shown in figure 8.9, by joining all points of equal concentration and extrapolating to zero angle, and then all points measured at equal angles and extrapolating these to zero concentration. For example, on the diagram the points corresponding to concentration c_3 are joined and extrapolated to intersect with an imaginary line corresponding to the value of $k'c_3$ on the abscissa; similarly all points measured at $90°$ are joined and extrapolated until the point corresponding to $\sin^2(90/2)$ is reached. This is done for each concentration and each angle and the extrapolated points are then lines of $\theta = 0$ and $c = 0$. Both lines, on extrapolation to the axis, should intersect at the same point. The intercept is then $(M_w)^{-1}$, the slope of the $\theta = 0$ line yields A_2, whereas $<S^2>$ is obtained from the initial slope s_i of the $c = 0$ line i.e.

$$<\dot{S}^2>_z = s_i M_w \, (3\lambda'^2/16\pi^2).\tag{8.29}$$

The radius of gyration calculated in this way for a polydisperse sample is a z-average.

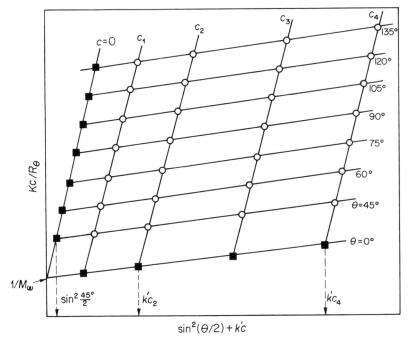

FIGURE 8.9. Typical Zimm plot showing the double extrapolation technique, where $-o-$ are the experimental points and $-\blacksquare-$ represent the extrapolated points.

8.10 Refractive index increment

Before M_w can be calculated from light scattering measurements, the specific refractive index increment $(d\bar{n}/dc)$ must be known for the particular polymer + solvent system under examination. It is defined as $(\bar{n} - \bar{n}_0)/c$ where $\bar{n}$ and $\bar{n}_0$ are the refractive indices of the solution and the solvent and c is the concentration. Measurements of $\Delta \bar{n} = (\bar{n} - \bar{n}_0)$ are made using a differential refractometer employing the same wavelength of light as used in the light scattering. The monochromatic beam (selected by filter) from a mercury vapour lamp is directed through a differential cell, consisting of a solution and solvent compartment separated by a diagonal glass wall. The deflection of the light beam is measured, first with solvent in the forward compartment and solution in the rear, giving deflection d_1; the position is reversed and deflection d_2 measured.

If similar readings for solvent alone, d_1^o and d_2^o, are obtained, then the total displacement Δd is

$$\Delta d = (d_1 - d_2) - (d_1^o - d_2^o). \qquad (8.30)$$

If the instrument is calibrated with aqueous KCl solutions of known $\Delta \bar{n}$ a relation,

$$\Delta \bar{n} = c' \Delta d, \qquad (8.31)$$

can be obtained where c' is the calibration constant. By measuring Δd for a number of concentrations of polymer, $\Delta \bar{n}$ is obtained from a knowledge of c', and $(d\bar{n}/dc)$ from the slope of the plot of $\Delta \bar{n}$ against c.

8.11 Small angle X-ray scattering

The theoretical outline presented for light scattering studies is valid for electromagnetic radiation of all wavelengths. For X-rays, λ is as low as 0.154 nm, and as this is much smaller than typical polymer dimensions structural information over small distances should be available from X-ray scattering. The intensity of scattering is a function of the electron density and therefore of the refractive index. The molar mass is then related to the excess electron density $\Delta \rho_e$ of solute over solvent for $\lambda = 0.154$ nm by

$$R_0 = (4.8 \text{ cm}^{-1}) M_w (\Delta \rho_e)^2 c, \qquad (8.32)$$

where R_0 is the Rayleigh ratio at $\theta = 0$. Experimental techniques are difficult because of the weak scattering, but the method has provided useful information on macromolecules with dimensions in the range 1 to 100 nm and, as such, is complementary to light scattering.

8.12 Ultracentrifuge

When macroscopic particles are allowed to settle in a liquid under gravity it is possible to determine their size and weight. Macromolecules in solution are usually much smaller and it would take years for them to overcome the Brownian motion and form a sediment. This problem can be overcome by subjecting them to an external force, strong enough to alter their spatial distribution by a significant amount in a short time. In 1925, Svedberg first achieved this by subjecting polymer solutions to large force fields, generated at high speeds of rotation.

The technique is now a well established method for measuring M_w and M_z for both synthetic and biological macromolecules and has the added advantage that measurements require only small quantities of material. The dilute solution of polymer is placed in a cell with a sector shaped centre piece in the form of a truncated cone, whose peak is located at the centre of the rotation. The shape ensures that convective disturbances are minimised during the transportation of molecules to the cell bottom. The cells are supported in a rotor of either titanium or aluminium alloy, which is attached to the drive motor by a fine steel wire, thereby allowing limited self-balancing to take place. The rotor is spun in a vacuum chamber to minimise frictional heating during high speed rotations, as speeds of up to 68 000 r.p.m., capable of producing 372 000 g can be generated. During rotation the cell passes through a collimated beam of light from a high pressure mercury lamp and the emergent beam then travels through the optical system to be recorded photographically. Three types of optical system are available, schlieren, interference, and u.v. absorption. Solvents, having densities and refractive indices sufficiently different from the polymer, are chosen to ensure movement of the polymer chains in the medium and the optical detection of this motion.

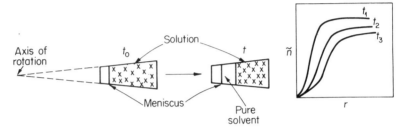

FIGURE 8.10. Schematic diagram of an ultracentrifuge cell showing boundary movement during a sedimentation run. The boundary movement can be followed by measuring the rapid change in refractive index $\tilde{n}$ on passing from solvent to solution.

Most commercial instruments are extremely versatile, with an extensive choice of rotor speeds and a temperature control system. Molar masses from 10^2 to 10^6 g mol^{-1} can be measured and this range is much wider than any other existing technique.

Two general methods are used to measure M, (1) sedimentation velocity and (2) sedimentation equilibrium.

SEDIMENTATION VELOCITY
The centrifuge is operated at high speeds to transport the polymer molecules through the solvent to the cell bottom if the solvent density is less than the polymer, or to the top (flotation) if the reverse is true. The rate of movement can be measured by following the change in refractive index $\tilde{n}$ through the boundary region. As the molecules sediment, a layer of pure solvent is left whose refractive index differs from the solution. The boundary is located by the sharp change in $\tilde{n}$ and its movement is followed as a function of time using one or other of the optical methods available.

If the centrifugal force moving the polymer through the solvent is F then

$$F = m(1 - \bar{v}_2\rho)r\omega^2, \tag{8.33}$$

where r is the distance between the boundary and the centre of rotation, $\bar{v}_2$ is the partial specific volume of the polymer, ω is the angular velocity, the mass of the molecule is $m = M/N_A$, and ρ is the density of the solution. This force will be balanced by the frictional resistance F of the medium for a particular velocity (dr/dt) and

$$F = 6\pi\eta R(dr/dt), \tag{8.34}$$

where R is the spherical radius of the polymer particle, and η is the viscosity of the medium. These two forces are in equilibrium when a uniform particle velocity is attained and

$$(M/N_A)(1 - \bar{v}_2\rho)r\omega^2 = 6\pi\eta R(dr/dt). \tag{8.35}$$

The steady-state velocity in a unit gravitational field can then be defined as the sedimentation constant S,

$$S = (1/\omega^2 r) \, (dr/dt), \tag{8.36}$$

and $\qquad S = (M/N_A)(1 - \bar{v}_2\rho)/6\pi\eta R = (M/N_A)(1 - \bar{v}_2\rho)/f \tag{8.37}$

where f is the frictional coefficient of the molecule and is related to the diffusion constant D by

$$D = kT/f \tag{8.38}$$

Substitution gives the Svedberg equation,

$$M_{SD} = \{RT/(1 - \bar{v}_2\rho)\}(S/D) \tag{8.39}$$

From this a molar mass M_{SD}, is calculated if both S and D are known. This average is close to M_w but is usually smaller and depends on the method used to measure D.

The term $(1 - \bar{v}_2\rho)$ is called the *buoyancy factor* and determines the direction of macromolecular transport in the cell. If the factor is positive, the polymer chains sediment away from the centre of rotation to the cell bottom, if negative, they move in the opposite direction and float to the top. The determination of M is absolute when S and D are known, but more commonly a relation of the form

$$S = K_s M^b, \tag{8.40}$$

is established, using polymer fractions of known M, for a given solvent + polymer system. This approach is similar to that used for the limiting viscosity number, which is a non-absolute method.

SEDIMENTATION EQUILIBRIUM
This technique is absolute, requiring no additional parameters. The calculation of M rests on a sound thermodynamic basis and relies on being able to measure the distribution of solute in the cell under equilibrium conditions. By using low rotation speeds, the molecules can diffuse and redistribute themselves in an

applied field in a manner dictated by their molar masses and the molar mass distribution.

The molar mass can be calculated by considering two points in the cell, distances r_1 and r_2 from the centre of rotation. At equilibrium the Gibbs free energy at each point is the same and $G_{r_1} = G_{r_2}$. The work required to transport a particle from r_1 to r_2 under a centrifugal force F is calculated from the change in the Gibbs free energy

$$\Delta G_S = -\int_{r_1}^{r_2} F \mathrm{d}r, \tag{8.41}$$

or
$$\Delta G_S = \left\{ -m(1 - \bar{v}_2 \rho) \omega^2 (r_2^2 - r_1^2)/2 \right\}. \tag{8.42}$$

This sedimentation process can be balanced by a back diffusion process and the Gibbs free energy change ΔG_D for this is

$$\Delta G_D = (RT/N_A) \ln(c_2/c_1), \tag{8.43}$$

where c_1 and c_2 are the polymer concentrations at r_1 and r_2. At equilibrium $\Delta G_S + \Delta G_D = 0$ and as $M = mN_A$ we have

$$M = 2RT \ln(c_2/c_1)/(1 - \bar{v}_2 \rho) \omega^2 (r_2^2 - r_1^2). \tag{8.44}$$

By calculating the concentrations at various points in the cell, a graph of log c against r^2 can be constructed and M_w follows from

$$M_w = \left\{ 2RT/(1 - \bar{v}_2 \rho) \omega^2 \right\} \left\{ 2.303 \mathrm{d} \log_{10} c / \mathrm{d}(r^2) \right\}. \tag{8.45}$$

This plot is normally curved for polydisperse samples and if $(M_w)_m$ is calculated from the limiting slope at the meniscus, while $(M_w)_b$ is the corresponding value at the cell bottom then

$$M_z = \left\{ (M_w)_b c_b - (M_w)_m c_m \right\} / (c_b - c_m), \tag{8.46}$$

where c_m and c_b are the respective concentrations.

The main disadvantage of the method lies in the long periods of time required to reach equilibrium. Several variations exist which reduce this time scale such as studying the approach to equilibrium, using short columns, or meniscus depletion techniques can be employed but all are outside the scope of this text.

The value of M_w calculated from equation (8.45) is, of course, an apparent value relating to the initial concentration of the solution, and extrapolation to zero concentration is necessary.

8.13 Viscosity
When a polymer dissolves in a liquid, the interaction of the two components stimulates an increase in polymer dimensions over that in the unsolvated state. Because of the vast difference in size between solvent and solute, the frictional properties of the solvent in the mixture are drastically altered, and an increase in

viscosity occurs which should reflect the size and shape of the dissolved solute, even in dilute solutions. This was first recognised in 1930 by Staudinger, who found that an empirical relation existed between the relative magnitude of the increase in viscosity and the molar mass of the polymer.

One of the simplest methods of examining this effect is by capillary viscometry. It has been shown that the ratio of the flow time of a polymer solution t to that of the pure solvent t_0 is effectively equal to the ratio of their viscosity (η/η_0) if the densities are equal. This latter approximation is reasonable for dilute solutions and provides a measure of the relative viscosity η_r.

$$\eta_r = (t/t_0) = (\eta/\eta_0) \qquad (8.47)$$

As this has a limiting value of unity, a more useful quantity is the specific viscosity

$$\eta_{sp} = \eta_r - 1 = (t - t_0)/t_0. \qquad (8.48)$$

Even in dilute solutions molecular interference is likely to occur and η_{sp} is extrapolated to zero concentration to obtain a measure of the influence of an isolated polymer coil. This is accomplished in either of two ways; η_{sp} can be expressed as a reduced quantity (η_{sp}/c) and extrapolated to $c = 0$ according to the relation

$$(\eta_{sp}/c) = [\eta] + k'[\eta]^2 c, \qquad (8.49)$$

and the intercept is the limiting viscosity number $[\eta]$ which is a characteristic parameter for the polymer in a particular solvent, k' is a shape dependent factor called the Huggins constant and has values between 0.3 and 0.9 for randomly coiling vinyl polymers. The alternative extrapolation method uses the inherent viscosity as

$$(\log \eta_r)/c = [\eta] + k''[\eta]^2 c, \qquad (8.50)$$

where k'' is another shape dependent factor. The dimensions of $[\eta]$ are the same as the reciprocal of the concentration.

When measuring $[\eta]$ solutions are filtered to remove spurious particles, then flow times for solvent and solutions are recorded in U-tube viscometers such as the "Cannon-Fenske" or the "Ubbelohde suspended level dilution" models. Dilution viscometers are most convenient when a concentration series is to be measured. In these the concentration can be changed *in situ*, whereas fresh solution concentrations of exactly the same volume must be introduced for each measurement in the non dilution Cannon-Fenske.

In the Ubbelohde viscometer an aliquot of solution of known volume is pipetted into bulb D through A. The solution is then pumped into E, by applying a pressure down A with C closed off; the pressure is released and C is opened to allow the excess solution to drain back into D. This leaves the end of the capillary open or suspended. Solution then flows down the capillary and drains round the sides of the bulb back into D, but as no back pressure from the excess solution exists, the volume in D plays no part in determining the flow time t. This suspended level is the feature which allows dilution to be carried out in D without affecting t. Thus addition of a known amount of solvent to the solution in D, followed by mixing, gives the next concentration in the series. The flow time t, is the time taken for the solution meniscus to pass from x to y in bulb E.

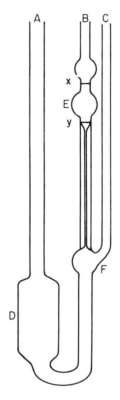

FIGURE 8.11. Suspended level dilution viscometer.

For a given polymer + solvent system at a specified temperature, $[\eta]$ can be related to M through the Mark-Houwink equation

$$[\eta] = K_v M^\nu, \tag{8.51}$$

and K_v and ν can be established, by calibrating with polymer fractions of known molar mass, and once this has been established for a system, $[\eta]$ alone will give M for an unknown fraction. This is normally achieved by plotting $\log [\eta]$ against $\log M$ and interpolation is then quite straightforward. Values of ν lie between 0.5 for a polymer dissolved in a theta-solvent to about 0.8 in very good solvents for linear randomly coiling vinyl polymers, and typical values for systems studied by viscosity and sedimentation are given in table 8.3. The exponents ν and b are indicative of solvent quality. When the solvent is ideal, *i.e.* a theta-solvent, both ν and b are 0.5, but as the solvent becomes thermodynamically better, and deviations from ideality larger, than ν increases and b decreases.

TABLE 8.3. A comparison of viscosity and sedimentation constants and exponents for several polymer + solvent systems from equations (8.51) and (8.40)

Polymer	Solvent	$\dfrac{T}{K}$	$\dfrac{10^2 K}{cm^3 g^{-1}}$	ν	$\dfrac{10^5 K_S}{s^{-1}}$	b
	Cyclohexene	298	1.63	0.68	3.85	0.42
Polystyrene	Chloroform	298	0.716	0.76	8.36	0.415
	Cyclohexane θ = 308	8.6	0.50	1.50	0.502
Poly(α-methylstyrene)	Cyclohexane θ = 310	7.8	0.50	1.86	0.50
	Toluene	310	1.0	0.72	4.02	0.43
Poly (vinyl acetate)	Butanone	298	4.2	0.62	9.8	0.38
Cellulose nitrate	Ethyl acetate	303	0.25	1.01	0.304	0.29
Cellulose	Cadoxen	298	250	0.75	19	0.40

8.14 Gel permeation chromatography

The determination of the molar mass distribution (MMD) by conventional fractionation techniques is time consuming, and a rapid, efficient, and reliable method, which allows the MMD to be determined in a matter of hours, has been developed. This is gel permeation chromatography (GPC) which separates polymer samples into fractions according to their molecular size, by means of a sieving action. This is achieved using a non-ionic stationary phase of packed spheres (often beads of crosslinked polystyrene) whose pore size distribution can be controlled. With a wide distribution of pore sizes in any support gel a separation into molecular size is obtained because the larger molecules dissolved in the solvent carrier, cannot diffuse into the pores, and are rapidly eluted, while the smaller ones penetrate further with decreasing size and are retarded correspondingly. Thus the large molecules leave the column first and the small ones last because they travel a much longer path.

A block diagram of a commercial instrument (Waters and Co.) is shown in figure 8.12. This is designed to fractionate small samples of polymer, and each fraction is detected by a differential refractometer system as it is eluted. Tetrahydrofuran is one of the commonly used solvents; these should be compatible with the stationary phase, have a similar polarity to prevent partitioning, and preferably have a high boiling temperature and low viscosity. Columns must be calibrated, and the elution times of samples of known M are measured for this purpose. As the technique appears to depend on the volume of the polymer, attempts to construct a universal calibration curve have been made. The most successful approach so far is to plot the product $[\eta]M$ against the elution volume V. Here $[\eta]M$ is proportional to the hydrodynamic volume V of

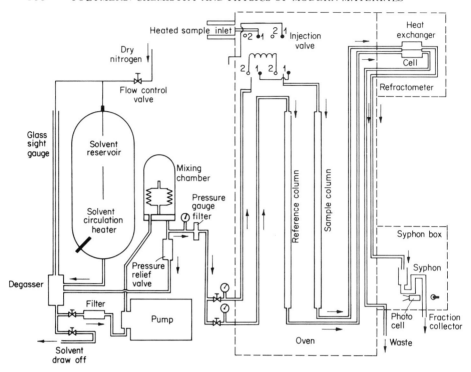

FIGURE 8.12. Diagram of a gel permeation chromatograph (Waters Associates Ltd.).

the polymer in solution, which is calculated from the modified Mark-Houwink relation $V = KM^{1 + \nu}$. An equation is obtained for standard polystyrene fractions of the form

$$\log M = a - bV, \qquad (8.52)$$

where a and b are constants, and if the Mark-Houwink equation is known for the particular polymer under study, the calibration curve can be directly related to the experimental curves via the Mark-Houwink constants. Analysis of the elution chromatogram also provides a measure of the heterogeneity ratio.

General Reading

P. W. Allen, *Techniques of Polymer Characterization.* Butterworths (1959).
N. M. Bikales, *Characterization of Polymers.* Wiley-Interscience (1971).
F. W. Billmeyer, *Textbook of Polymer Science,* Chapter 3. John Wiley and Sons (1962).
R. D. Bonnar, M. Dimbat, and F. H. Stross, *Number Average Molecular Weights.* Interscience (1958).
T. J. Bowen, *An Introduction to Ultracentrifugation.* Wiley Interscience (1970).

B. Carroll, *Physical Methods in Macromolecular Chemistry*, Vol. 2. Marcel Dekker (1972).

M. B. Huglin, *Light Scattering from Polymer Solutions.* Academic Press (1972).

J. F. Johnson and R. F. Porter, *Analytical Gel Permeation Chromatography.* John Wiley and Sons (1968).

D. Margerison and G. C. East, *Introduction to Polymer Chemistry*, Chapter 2. Pergamon Press (1967).

H. Morawetz, *Macromolecules in solution.* Interscience Publishers Inc. (1965).

D. A. Smith, *Addition Polymers*, Chapter 5. Butterworths (1968).

References

1. T. G. Fox, J. B. Kinsinger, H. F. Mason and E. M. Schuele, *Polymer,* **3**, 71 (1962).

CHAPTER 9

Polymer Characterization—Chain Dimensions and Structures

As the size and shape of a polymer chain are of considerable interest to the polymer scientist it is useful to know how these factors can be assessed. Much of the information can be derived from studies of dilute solutions; an absolute measurement of polymer chain size can be obtained from light scattering, when the polymer is large compared with the wavelength of the incident light. Sometimes the absolute measurement cannot be used but the size can be deduced indirectly from viscosity measurements, which are related to the volume occupied by the chain in solution. Armed with this information we must now determine how meaningful it is and to do this a clearer understanding of the factors governing the shape of the polymer is required. We can confine ourselves to models of the random coil, as this is usually believed to be most appropriate for synthetic polymers; other models – rods, discs, spheres, spheroids – are also postulated, but need not concern us at this level.

9.1 Average chain dimensions

A polymer chain in dilute solution can be pictured as a coil, continuously changing its shape under the action of random thermal motions. This means, that at any time, the volume occupied by a chain in solution, could differ from that occupied by its neighbours, and these size differences are further accentuated by the fact that each sample will contain a variety of chain lengths. Taking these two points into consideration leads us to the conclusion that meaningful chain dimensions can only be values averaged over the many conformations assumed. Two such averages have been defined: (a) the average root mean square distance between the chain ends $\langle \bar{r}^2 \rangle^{1/2}$; and (b) the average root mean square radius of gyration $\langle \bar{S}^2 \rangle^{1/2}$ which is a measure of the average distance of a chain element from the centre of gravity of the coil. The angular brackets denote averaging due to chain polydispersity in the sample and the bar indicates averaging for the many conformational sizes available to chains of the same molar mass.

The two quantities are related, in the absence of excluded volume effects, for simple chains by

$$\langle \bar{r}^2 \rangle^{1/2} = \langle 6 \bar{S}^2 \rangle^{1/2}, \qquad (9.1)$$

but as the actual dimensions obtained can depend on the conditions of the measurement, other factors must also be considered.

9.2 Freely-jointed chain model

The initial attempts to arrive at a theoretical representation of the dimensions of a linear chain, treated the molecule as a number n of chain elements, joined by bonds of length l. By assuming the bonds act like universal joints, complete freedom of rotation about the chain bonds can be postulated. This model allows the chain to be pictured as in figure 9.1(a) which resembles the path of a diffusing gas molecule and as random flight statistics have proved useful in describing gases, a similar approach is used here. In two dimensions the diagram is more picturesquely called the "drunkard's walk" and r_f is estimated by considering first the simplest case of two links. The end-to-end distance r_f follows from the cosine law that

$$OB^2 = OA^2 + AB^2 - 2(OA)(AB)\cos \theta \qquad (9.2)$$

see figure 9.1(b), or

$$r_f^2 = 2l^2 - 2l^2 \cos \theta. \qquad (9.3)$$

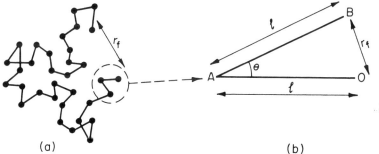

(a)   (b)

FIGURE 9.1. (a) Random walk chain of 32 steps, length l and (b) cosine law for two bonds.

When n is large, the angle θ will vary over all possible values so that the sum of all these terms will be zero, and as $\cos \theta = -\cos(\theta + \pi)$ equation (9.3) will reduce to

$$r_f^2 = nl^2. \qquad (9.4)$$

This shows that the distance between the chain ends, for this model is proportional to the square root of the number of bonds and so, is considerably shorter than a fully extended chain.

The result is the same if the molecule is thought to occupy three-dimensional space, but if it is centred on a co-ordinate system both positive and negative contributions occur with equal probability. To overcome this the dimension is expressed always as the square which eliminates negative signs. This model is however unrealistic. Polymer chains occupy a volume in space, and the dimensions of any macromolecule are influenced by the bond angles and by interactions between the chain elements. These interactions can be classified into two groups: (i) *Short range* interactions which occur between neighbouring atoms or groups, and are usually forces of steric repulsion caused by the overlapping of electron clouds; (ii) *Long range* interactions which are comprised of attractive and repulsive forces between segments, widely separated in a chain, that occasionally approach one another during molecular flexing, and between segments and solvent molecules. These are often termed *excluded volume* effects.

9.3 Short range effects
The expansion of a covalently bonded polymer chain will be restricted by the valance angles between each chain atom. In general this angle is θ for a homo-atomic chain and equation (9.4) can be modified to allow for these short range interactions.

$$\langle \bar{r}^2 \rangle_{of} = nl^2 \, (1 - \cos\theta)/(1 + \cos\theta) \tag{9.5}$$

For the simplest case of an all carbon backbone chain such as polyethylene, $\theta \approx 109°$ and $\cos\theta = -\frac{1}{3}$ so that equation (9.5) reduces to

$$\langle \bar{r}^2 \rangle_{of} = 2nl^2. \tag{9.6}$$

This indicates that the polyethylene chain is twice as extended as the freely jointed chain model when short range interactions are considered.

9.4 Chain stiffness
As we have already seen in chapter 1 for butane and polyethylene, steric repulsions impose restrictions to bond rotation. This means that equation (9.5) has to be modified further and now becomes

$$\langle r^2 \rangle_{o} = nl^2 \, \frac{(1 - \cos\theta)}{(1 + \cos\theta)} \cdot \frac{(1 - \langle\cos\phi\rangle)}{(1 + \langle\cos\phi\rangle)} \tag{9.7}$$

where $\langle\cos\phi\rangle$ is the average cosine of the angle of rotation of the bonds in the backbone chain. The parameter $\langle \bar{r}^2 \rangle_{o}$ is the average mean square of the *unperturbed dimension*, which is a characteristic parameter for a given polymer chain.

The freely jointed dimensions are now more realistic when restricted by the factor ζ the skeletal factor – composed of the two terms

$$\zeta = \sigma^2 (1 - \cos\theta)/(1 + \cos\theta), \tag{9.8}$$

where σ is known as the steric parameter and is $(1 - \langle\cos\phi\rangle)/(1 + \langle\cos\phi\rangle)$ for simple chains.

For more complex chains, containing rings or heteroatomic chains, e.g. polydienes, polyethers, polysaccharides, and proteins, an estimate of σ is obtained from

$$\sigma^2 = \langle\bar{r}^2\rangle_0 / \langle\bar{r}^2\rangle_{of}. \tag{9.9}$$

Values of the unperturbed dimension can be obtained experimentally from dilute solution measurements made either directly in a theta-solvent (see section 8.9) or by using indirect measurements in non-ideal solvents and employing an extrapolation procedure. The geometry of each chain allows the calculation of $\langle\bar{r}^2\rangle_{of}$ and results are expressed either as σ or as the characteristic ratio $\{\langle\bar{r}^2\rangle_0/nl^2\}$. Both provide a measure of chain stiffness in dilute solution. The range of values normally found for σ is from about 1.5 to 2.5 as shown in table 9.1.

TABLE 9.1. Chain stiffness parameters and typical dimensions

Polymer	$T/\overset{\circ}{K}$	σ	$[\langle\bar{r}^2\rangle_0/nl^2]$
Polypropylene (isotactic)	408	1.53	4.67
Polypropylene (atactic)	408	1.65	5.44
Natural rubber	293	1.67	4.70
Guttapercha	333	1.38	7.35
Polystyrene	308	2.23	10.00
Poly(methyl methacrylate)			
(Isotactic)	298	2.28	10.40
(Atactic)	298	2.01	8.10
(Syndiotactic)	308	1.94	7.50

9.5 Treatment of dilute solution data

We can now examine some of the ways of calculating the polymer dimensions from experimental data.

THE SECOND VIRIAL COEFFICIENT

An investigation of the dilute solution behaviour of a polymer can provide useful information about the size and shape of the coil, the extent of polymer-solvent interaction and the molar mass. Deviations from ideality, as we have seen in section 8.7, are conveniently expressed in terms of virial expansions, and when solutions are sufficiently dilute, the results can be adequately described by the terms up to the second virial coefficient A_2 while neglecting higher terms.

The value of A_2 is a measure of solvent-polymer compatibility, as the parameter reflects the tendency of a polymer segment to exclude its neighbours from the volume it occupies. Thus a large positive A_2 indicates a good solvent for the polymer while a low value (sometimes even negative) shows that the solvent is relatively poor. The virial coefficient can be related to the Flory dilute solution parameters by

$$A_2 = \psi_1(1 - \Theta/T)(\bar{v}_2^2/V_1)F(x) \tag{9.10}$$

where $F(x)$ is a molar mass dependent function of the excluded volume. The exact form of $F(x)$ can be defined explicitly by one of several theories, and while each leads to a slightly different form, all predict that $F(x)$ is unity when theta conditions are attained and the excluded volume effect vanishes. Equation (9.10) can be used to analyse data such as that in figure 7.6. Once Θ has been located, the entropy parameter ψ_1 can be calculated by replotting the data as $\psi_1 F(x)$ against T. Extrapolation to $T = \Theta$, where $F(x) = 1$, allows ψ_1 to be estimated for the system under theta conditions. This method of measuring Θ and ψ_1 is only accurate when the solvent is poor, and extrapolations are short.

The dependence of A_2 on M, which is foreshadowed in equation (7.9), can often be predicted, for good solvents, by a simple equation

$$A_2 = kM^{-\gamma}, \tag{9.11}$$

where γ varies from 0.15 to 0.4, depending on the system and k is a constant.

EXPANSION FACTOR α

The value of A_2 will tell us whether or not the size of the polymer coil, which is dissolved in a particular solvent, will be perturbed or expanded over that of the unperturbed state, but the extent of this expansion is best estimated by calculating the expansion factor α.

If the temperature of a system, containing a polymer of finite M, drops much below Θ the number of polymer-polymer contacts increases until precipitation of the polymer occurs. Above this temperature, the chains are expanded, or perturbed, from the equilibrium size attained under pseudo-ideal conditions, by long range interactions. The extent of this coil expansion is determined by two long range effects. The first results from the physical exclusion of one polymer segment by another from a hypothetical lattice site which reduces the number of possible conformations available to the chain. This serves to lower the probability that tightly coiled conformations will be favoured. The second is observed in very good solvents, where the tendency is for polymer-solvent interactions to predominate, and leads to a preference for even more extended conformations. In a given solvent an equilibrium conformation is eventually achieved when the forces of expansion are balanced by forces of contraction in the molecule. The tendency to contract arises from both the polymer-polymer interactions and the resistance to expansion of the chain into over extended and energetically less favoured conformations.

The extent of this coil perturbation by long range effects is measured by an expansion factor α, introduced by Flory. This relates the perturbed and unperturbed dimensions by

$$\langle \bar{S}^2 \rangle^{1/2} = \alpha \langle \bar{S}^2 \rangle_0^{1/2} \tag{9.12}$$

In good solvents (large, positive A_2) the coil is more extended than in poor solvents (low A_2) and α is correspondingly larger. Since α is solvent and temperature dependent a more characteristic dimension to measure for the polymer is $\langle \bar{S}^2 \rangle_0^{1/2}$, which can be calculated from light scattering in a theta solvent, or indirectly as next described.

FLORY-FOX THEORY

The molecular dimensions of a polymer chain in any solvent can be calculated directly from light scattering measurements, using equation (8.29), if the coil is large enough to scatter light in an asymmetric manner, but when the chain is too short to be measured accurately in this way an alternative technique has to be used.

Flory and Fox suggested that as the viscosity of a polymer solution will depend on the volume occupied by the polymer chain, it should be feasible to relate coil size and $[\eta]$. They assumed that if the unperturbed polymer is approximated by a hydrodynamic sphere, then $[\eta]_\theta$, the limiting viscosity number in a theta solvent, could be related to the square root of the molar mass by

$$[\eta]_\theta = K_\theta M^{1/2}, \tag{9.13}$$

where $\qquad K_\theta = \Phi(\bar{r}_0^2/M)^{3/2}. \tag{9.14}$

Equations (9.13) and (9.14) are actually derived for monodisperse samples, and when measurements are performed with heterodisperse polymers, the appropriate averages to use are M_n and $\langle \bar{r}^2 \rangle_{on}$. The parameter Φ was originally considered to be a universal constant, but experimental work suggests that it is a function of the solvent, molar mass, and heterogeneity. Values can vary from an experimental one of 2.1×10^{23} to a theoretical limit of about 2.84×10^{23} when $[\eta]$ is expressed in cm^3 g^{-1}. A most probable value of 2.5×10^{23} has been found to be acceptable for most flexible heterodisperse polymers in good solvents.

For non-ideal solvents equation (9.13) can be expanded to give

$$[\eta] = K_\theta M^{1/2} \alpha_\eta^3, \tag{9.15}$$

where $\alpha_\eta^3 = [\eta]/[\eta]_\theta$ is the linear expansion factor, pertaining to viscosity measurements, and is a measure of long range interactions. As the derivation is based on an unrealistic Gaussian distribution of segments in good solvents, it has been suggested that α_η is related to the more direct measurement of α in equation (9.12) by

$$\alpha_\eta^3 = \alpha^{2.43}. \tag{9.16}$$

Considerable experimental evidence exists to support this conclusion.

INDIRECT ESTIMATES OF $\langle \bar{r}^2 \rangle_0^{1/2}$

It is not always possible to find a suitable theta-solvent for a polymer and methods have been developed which allow unperturbed dimensions to be estimated in non-ideal (good) solvents.

Several methods of extrapolating data for $[\eta]$ have been suggested. The most useful of these was proposed by Stockmayer and Fixman, using the equation:

$$[\eta]M^{-1/2} = K_\theta + 0.51\ \Phi B' M^{1/2}, \tag{9.17}$$

where Φ is assumed to adopt its limiting theoretical value, B' is related to the thermodynamic interaction parameter χ_1 by

$$B' = \bar{v}_2^2(1 - 2\chi_1)/V_1 N_A, \tag{9.18}$$

and examination of equation (9.10) shows that B' is also proportional to A_2. The unperturbed dimension can be estimated by plotting $[\eta]M^{-1/2}$ against $M^{1/2}$; K_θ is obtained from the intercept and $\langle \bar{r}^2 \rangle_0$ is calculated from equation (9.14).

A similar procedure has been proposed by Cowie and Bywater, in which the intrinsic frictional coefficient $[f]$ measured from sedimentation or diffusion experiments, will provide the same information using

$$[f]\ M^{1/2} = K_f + 0.201\ K_f^{-2}P_0^3\ B'M^{1/2},\ \ldots \tag{9.19}$$

where

$$K_f = P_0[\langle \bar{r}_0^2 \rangle/M]^{1/2}$$

and P_0 is a "constant" with a limiting value of 5.2.

These extrapolation procedures all depend on the validity of the theoretical treatment and reliability must be judged in this light. Fortunately, it has been demonstrated that most non-polar polymers can be treated in this way and results agree well with direct measurements of $\langle \bar{r}^2 \rangle_0^{1/2}$. For more polar polymers, specific solvent effects become more pronounced and extrapolations have to be regarded with corresponding caution.

INFLUENCE OF TACTICITY ON CHAIN DIMENSIONS

Studies of the dilute solution behaviour of polymers with a specific stereostructure have revealed that the unperturbed dimensions may depend on the chain configuration. This can be seen from the data in table 9.1 where isotactic, syndiotactic, and atactic poly(methyl methacrylate) have different σ values. If the size of a polymer chain can be affected by its configuration, the microstructure must be well characterized before an accurate assessment of experimental data can be made. This can be achieved using n.m.r. and infrared techniques.

9.6 Nuclear magnetic resonance (n.m.r.)

High resolution n.m.r. has proved to be a particularly useful tool in the study of the microstructure of polymers in solution, where the extensive molecular motion reduces the effect of long range interactions and allows the short range effects to dominate. Interpretation of chain tacticity, based on the work of Bovey and Tiers, can be illustrated using poly(methyl methacrylate). The three possible steric configurations are shown in figure 9.2 where R is the group $-COOCH_3$.

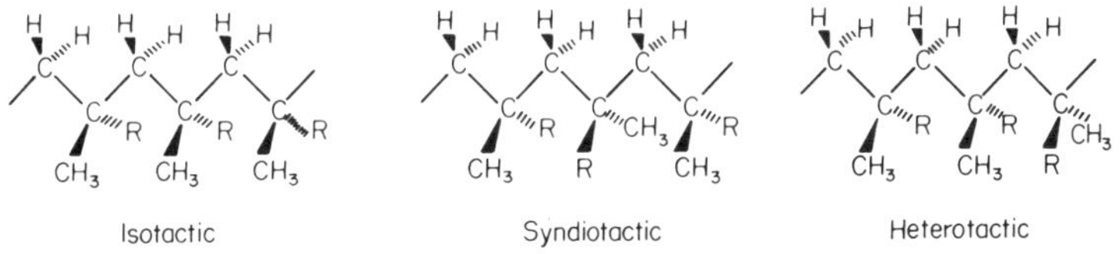

Isotactic   Syndiotactic   Heterotactic

FIGURE 9.2. Stereoregular triads for poly(methyl methacrylate) where $R = -COOCH_3$.

For the purposes of n.m.r. measurements three consecutive monomer units in a chain are considered to define a configuration and called a triad. The term heterotactic is used now to define a triad which is neither isotactic nor syndiotactic. In the structures shown, the three equivalent protons of the α-methyl group absorb radiation at a single frequency, but this frequency will be different for each of the three kinds of triad, because the environment of the α-methyl groups in each is different. For poly(methyl methacrylate) samples, which were prepared under different conditions to give the three forms, resonances at $\tau = 8.78$, 8.95 and 9.09 were observed, which were assigned to the isotactic, heterotactic, and syndiotactic triads respectively. Thus in a sample with a mixture of configurations a triple peak will be observed and the area under each of these peaks will correspond to the amount of each triad present in the polymer chain. This is illustrated in figure 9.3, where one sample is predominantly isotactic, but also contains smaller percentages of the heterotactic and syndiotactic configurations.

The analysis can be carried further. The fraction of each configuration, P_i, P_h, and P_s, measured from the respective peak areas, can be related to p_m the probability that a monomer adding on to the end of a growing chain will have the same configuration as the unit it is joining. This leads to the relations

$$P_i = p_m^2, P_s = (1 - p_m)^2, \quad \text{and} \quad P_h = 2p_m(1 - p_m).$$

Curves plotted according to this simple analysis are shown in figure 9.4 where they are compared with experimental data obtained for various tactic forms of poly(α-methyl styrene).

Differences in the microstructure of polydienes and copolymers can also be made using n.m.r. In the polydienes the difference between 1,2- and 1,4-addition

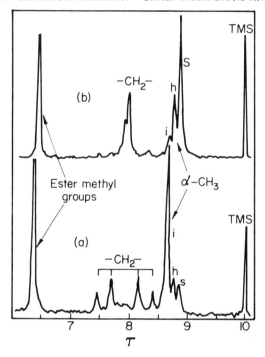

FIGURE 9.3. The n.m.r. spectra for (a) an isotactic sample and (b) a predominantly syndiotactic sample of poly(methyl methacrylate).

can be distinguished on examination of the resonance peaks corresponding to terminal olefinic protons, found at $\tau = 4.9$ to 5.0, and non-terminal olefinic protons observed at $\tau = 4.6$ to 4.7.

Not only is the local field acting on the nucleus altered by environment, it is also sensitive to molecular motion, and it has been observed that as the molecular motion within a sample increases, the resonance lines become narrower. Determination of the width, or second moment, of an n.m.r. resonance line, then provides a sensitive measure of low frequency internal motions in solid polymers and can be used to study transitions and segmental rotations in the polymer sample. Line widths are also altered by the polymer crystallinity. Partially crystalline polymers present complex spectra as they are multi-phase materials, in which the molecular motions are more restricted in the crystalline phase than in the amorphous phase. However, attempts to estimate percentage crystallinity in a sample using n.m.r. have not been particularly successful. The method is illustrated in figure 9.5 for poly(tetrafluoroethylene) where glass and other transitions are readily detected. Below 200 K the chains are virtually immobile, but above 200 K the lines sharpen as — CF_2 — rotation begins. This is associated with the glass transition, but the way the line width increases in this region is governed by sample crystallinity.

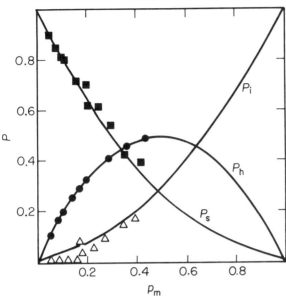

FIGURE 9.4. Theoretical curves for P as a function of p_m for each of the three configurations. Points represent experimental data for poly(α-methyl styrene) and illustrate the validity of the analysis. (From data by Brownstein, Bywater, and Worsfold (1961).)

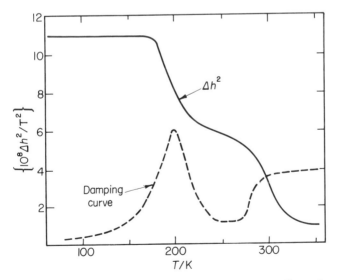

FIGURE 9.5. The n.m.r. line width Δh (the unit is the tesla T) as a function of temperature for poly(tetrafluoroethylene). The mechanical damping curve (----) is included for comparison. (Adapted from Sauer and Woodward (1960).)

9.7 Infrared spectroscopy

Infrared spectroscopy can be used to characterize long chain polymers because the infrared active groups, present along the chain, absorb as if each was a localized group in a simple molecule. Identification of polymer samples can be made by making use of the "finger-print" region, where it is least likely for one polymer to exhibit exactly the same spectrum as another. This region lies within the range 6.67 to 12.50 μm.

In addition to identification, the technique has been used to elucidate certain aspects of polymer microstructure, such as branching, crystallinity, tacticity, and *cis-trans* isomerism. The relative proportions of *cis*-1,4-, *trans*-1,4-, and 1,2- addition in polybutadienes can be ascertained by making use of the differences in absorption between (CH) out of plane bending vibrations, which depend on the type of substitution at the olefinic bond. Terminal and internal groups can also be distinguished, as an absorption band at about 11.0 μm is characteristic of a vinyl group and indicates 1,2-addition. The *cis*-1,4-addition is characterized by an absorption band at about 13.6 μm, whereas the *trans*-1,4 configuration exhibits a band at about 10.4 μm. An estimate of *cis-trans* isomerism can be made by measuring the absorbance A of each band, where $A = \log_{10}(I_0/I)$ and I_0 and I are the intensities of the incident and transmitted radiation respectively. This is calculated by locating a base line across the minima on either side of the absorption band and the vertical height to the top of the band from the base line is converted into a composition using the equation

$$P_{cis} = 3.65\, A_{cis}/(3.65\, A_{cis} + A_{trans}),$$

where P_{cis} is the fraction of *cis* configuration, A_{cis} is the absorbance at 13.6 μm, A_{trans} the absorbance at 10.4 μm, and if we assume that the 1,2 content is negligible. Polyisoprenes can also be analyzed in this way, only now the bands at 11.0 and 11.25 μm are used to estimate the 1,2- and 3,4-addition, while a band at 8.7 μm corresponds to the *trans*-1,4 linkage.

The infrared spectra of highly stereoregular polymers are distinguishable from those of their less regular counterparts, but many of the differenct s can be attributed to crystallinity rather than tacticity as such. The application of infrared to stereostructure determination in polymers is less reliable than n.m.r., but has achieved moderate success for poly(methyl methacrylate) and poly- propylene. In poly(methyl methacrylate) a methyl deformation at 7.25 μm, is unaffected by microstructure, and comparison of this with a band at 9.40 μm, which is present only in atactic or syndiotactic polymers allows an estimate of the syndiotacticity to be made from the ratio $\{A(9.40\,\mu\text{m})/A(7.25\,\mu\text{m})\}$. Similarly $\{A(6.75\,\mu\text{m})/A(7.25\,\mu\text{m})\}$ provides a measure of the isotactic content. An alternative method is to calculate the quantity J as an average of the two equations

$$J_1 = 179\{A(9.40\,\mu\text{m})/A(10.10\,\mu\text{m})\} + 27$$

$$J_2 = 81.4\{A(6.75\,\mu\text{m})/A(7.25\,\mu\text{m})\} - 43$$

where the absorption band at $10.10\,\mu$m is now used. If J lies between 100 and 115 a highly syndiotactic polymer is indicated, if between 25 and 30 the polymer is highly isotactic. For polypropylene, the characteristic band for the syndio-tactic polymer appears at $11.53\,\mu$m, and the syndiotactic index I_s is $2A(11.53\,\mu\text{m})/\{A(2.32\,\mu\text{m}) + A(2.35\,\mu\text{m})\}$. Values of I_s about 0.8 indicate highly syndiotactic samples. Spectra can be measured in a number of ways; for soluble polymers a film can be cast, perhaps even on the NaCl plate to be used and examined directly. Measurements can also be made in solution, if the solvent absorption in any important region is low, or by a differential method.

9.8 X-ray diffraction

The extent of sample crystallinity can influence the behaviour of a polymer sample greatly. A particularly effective way of examining partially crystalline polymers is by X-ray diffraction. The crystallites present in a powdered or unoriented polymer sample diffract X-ray beams from parallel planes for incident angles θ which are determined by the Bragg equation

$$n\lambda = 2d \sin \theta, \tag{9.21}$$

where λ is the wavelength of the radiation, d is the distance between the parallel planes in the crystallites, and n is an integer. The reinforced waves reflected by all the small crystallites produce diffraction rings, or haloes, which are sharply defined for highly crystalline materials and become increasingly diffuse when the amorphous content is high.

If the polymer sample is oriented, by drawing a fibre, or by applying tension to a film, the crystallites tend to become aligned in the direction of the stress and the X-ray pattern is improved. In some samples of stereoregular or symmetrical polymers, the degree of three-dimensional ordering of the chains may be sufficiently high to allow a structural analysis of the polymer to be accomplished.

Sample crystallinity can be estimated from the X-ray patterns by plotting the density of the scattered beam against the angle of incidence. If this can be done for an amorphous sample and a corresponding sample which is highly crystalline, a relative measure of crystallinity for other samples of the same polymer can be obtained. In figure 9.6 the shaded portion is the amorphous polypropylene, while the maxima arise from the crystallites.

9.9 Thermal analysis

When a substance undergoes a physical or chemical change a corresponding change in enthalpy is observed. This forms the basis of the technique known as differential thermal analysis (DTA) in which the change is detected by measuring the enthalpy difference between the material under study and an inert standard.

The sample is placed in a heating block and warmed at a uniform rate. The sample temperature is then monitored by means of a thermocouple and compared with the temperature of an inert reference such as powdered alumina, or simply an empty sample pan, which is subjected to the same linear heating

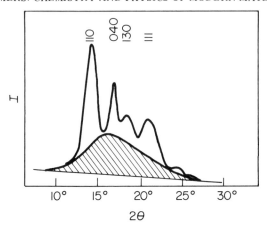

FIGURE 9.6. X-ray diffraction curves; the intensity I as a function of angle for totally amorphous polypropylene (shaded area), and for a sample with a 50 per cent crystalline content.

programme. As the temperature of the block is raised at a constant rate (5 to 20 K min^{-1}) the sample temperature T_s and that of the reference T_r will keep pace until a change in the sample takes place. If the change is exothermic T_s will exceed T_r for a short period, but if it is endothermic T_s will temporarily lag behind T_r. This temperature difference ΔT is recorded and transmitted to a chart recorder where changes such as melting or crystallization are recorded as peaks. A third type of change can be detected. Since the heat capacities of sample and reference are different ΔT is never actually zero, and a change in heat capacity, such as that associated with a glass transition, will cause a shift in the base line. All three possibilities are shown in figure 9.7 for quenched terylene.

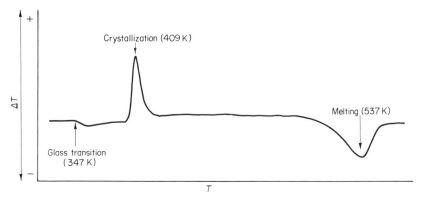

FIGURE 9.7. A DTA curve for quenched terylene showing the glass transition, melting endotherm, and a crystallization exotherm.

Other changes such as sample decomposition, crosslinking, and the existence of polymorphic forms can also be detected. As ΔT measured in DTA is a function of the thermal conductivity and bulk density of the sample, it is non-quantitative and relatively uninformative. To overcome these drawbacks an alternative procedure known as differential scanning calorimetry (DSC) is used. This technique retains the constant mean heat input but instead of measuring the temperature difference during a change, a servo-system immediately increases the energy input to either sample or reference to maintain both at the same temperature. The thermograms obtained are similar to DTA, but actually represent the amount of electrical energy supplied to the system, not ΔT, and so the areas under the peaks will be proportional to the change in enthalpy which occurred. An actual reference sample can be dispensed with in practice and an empty sample pan used instead. Calibration of the instrument will allow the heat capacity of a sample to be calculated in a quantitative manner. This information is additional to that gained on crystallization, melting, glass transitions, and decompositions.

General Reading

N. M. Bikales, *Characterization of Polymers.* Wiley-Interscience (1971).
T. M. Birshtein and D. B. Ptitsyn, *Conformations of Macromolecules.* Interscience Inc. (1966).
F. A. Bovey, *Polymer Conformation and Configuration.* Academic Press (1969).
B. Carroll, *Physical Methods in Macromolecular Chemistry,* Vol. 2. Marcel Dekker (1972).
A. Elliott, *Infrared Spectra and Structure of Organic Long Chain Polymers.* Edward Arnold (1969).
P. J. Flory, *Principles of Polymer Chemistry,* Chapters 10 and 14. Cornell Univ. Press, Ithaca, N.Y. (1953).
P. J. Flory, *Statistical Mechanics of Chain Molecules.* Interscience Publishers Inc. (1969).
J. F. Johnson and R. F. Porter, *Analytical Calorimetry.* Plenum Press (1968).
B. Ke, *Newer Methods of Polymer Characterization.* Interscience Publishers Inc. (1964).
D. A. Smith, *Addition Polymers,* Chapter 7. Butterworths (1968).
A. V. Tobolsky and H. Mark, *Polymer Science and Materials,* Chapter 3. Wiley-Interscience (1971).
M. V. Volkenstein, *Configurational Statistics of Polymeric Chains.* Interscience Publishers Inc. (1963).

References

1. S. Brownstein, S. Bywater, and D. J. Worsfold, *Makromol. Chem.,* 58, 127 (1961).
2. J. A. Sauer and A. E. Woodward, *Reviews in Modern Physics,* 32, 88 (1960).

CHAPTER 10

The Crystalline State

10.1 Introduction

When polymers are irradiated by a beam of X-rays, scattering produces diffuse haloes on the photographic plate for some polymers, while for others a series of sharply defined rings superimposed on a diffuse background is recorded. The former are characteristic of amorphous polymers, and illustrate that a limited amount of short range order exists in most polymeric solids. The latter patterns are indicative of considerable three-dimensional order and are typical of poly-crystalline samples containing a large number of unoriented crystallites associated with amorphous regions. The rings are observed to sharpen into arcs, or discrete spots, if the polymer is drawn or stretched, a process which orients the axes of the crystallites in one direction.

The occurrence of significant crystallinity in a polymer sample is of considerable consequence to a materials scientist. The properties of the sample − the density, optical clarity, modulus, and general mechanical response − all change dramatically when crystallites are present and the polymer is no longer subject to the rules of linear visco-elasticity, which apply to amorphous polymers as outlined in Chapter 12. However, a polymer sample is rarely completely crystalline and the properties also depend on the amount of crystalline order.

It is important then to examine crystallinity in polymers and determine the factors which control the extent of crystallinity.

10.2 Mechanism of crystallization

A polymer in very dilute solution can be effectively regarded as an isolated chain whose shape is governed by short and long range inter- and intra-molecular interactions. In the aggregated state this is no longer true, the behaviour of the chain is now influenced largely by the proximity of the neighbouring chains and the secondary valence forces which act between them. These factors determine the orientation of chains relative to each other in the undiluted state, and this is essentially an interplay between the entropy and internal energy of the system which is expressed in the usual thermodynamic form

$$G = (U + pV) - TS.$$

In the melt, polymers normally attain a state of maximum entropy consistent with a stable state of minimum free energy. Crystallization is a process involving the orderly arrangement of chains and is consequently associated with a large negative entropy of activation. If a favourable free energy change is to be obtained for crystallite formation, the entropy term has to be offset by a large negative energy contribution.

The alignment of polymer chains at specific distances from one another to form crystalline nuclei will be assisted when intermolecular forces are strong. The greater this interaction between chains the more favourable will be the energy parameter and this provides some indication of the type of chain which might be expected to crystallize from the melt, *viz.*

(1) Symmetrical chains which allow the regular close packing required for crystallite formation.
(2) Chains possessing groups which encourage strong intermolecular attraction thereby stabilizing the alignment.

In addition to the thermodynamic requirements, kinetic factors relating to the flexibility and mobility of a chain in the melt must also be considered. Thus polyisobutylene $+ CH_2C(CH_3)_2 +_n$ might be expected to crystallize because the chain is symmetrical, but it will only do so if maintained at an optimum temperature for several months. This is presumably a result of the flexibility of the chain which allows extensive convolution thereby impeding stabilization of the required long range alignment.

The creation of a three-dimensional ordered phase from a disordered state is a two stage process. Just above its melting temperature a polymer behaves like a highly viscous liquid in which the chains are all tangled up with their neighbours. Each chain pervades a given volume in the sample, but as the temperature decreases the volume available to the molecule also decreases. This in turn restricts the number of disordered conformational states available to the chain due to the constraining influence of intramolecular interactions among chains in juxtaposition. As a result there is an increasing tendency for the polymer to assume an ordered conformation in which the chain bonds are in the rotational states of lowest energy. However, various other factors will tend to oppose crystallization; chain entanglements will hinder the diffusion of chains into suitable orientations and if the temperature is above the melting temperature, thermal motions will be sufficient to disrupt the potential nuclei before significant growth can take place. This restricts crystallization to a range of temperatures between T_g and T_m.

The first step in crystallite formation is the creation of a stable nucleus

brought about by the ordering of chains in a parallel array, stimulated by
intramolecular forces, followed by the stabilization of long range order by the
secondary valence forces which aid the packing of molecules into a three-
dimensional ordered structure.

The second stage is the growth of the crystalline region, the size of which is
governed by the rate of addition of other chains to the nucleus. As this growth
is counteracted by thermal redispersion of the chains at the crystal-melt
interface, the temperature must be low enough to ensure that this disordering
process is minimal.

10.3 Temperature and growth rate

Measureable rates of crystallization occur between $(T_m - 10$ K$)$ and $(T_g + 30$ K$)$,
a range in which the thermal motion of the polymer chains is conducive to the
formation of stable ordered regions. The growth rate of crystalline areas passes
through a maximum in this range as illustrated in figure 10.1 for isotactic
polystyrene. Close to T_m the segmental motion is too great to allow many
stable nuclei to form, while near T_g the melt is so viscous that molecular motion
is extremely slow.

As the temperature drops from T_m, the melt viscosity, which is a function
of the molar mass, increases and the diffusion rate decreases, thereby giving the
chains greater opportunity to rearrange themselves to form a nucleus. This means

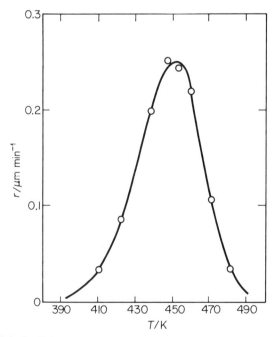

FIGURE 10.1. Radial growth rate r of spherulites of isotactic polystyrene as a
function of the crystallization temperature.

that there will exist an optimum temperature of crystallization, which depends largely on the interval T_m to T_g, but also on the molar mass of the sample.

The melt usually has to be supercooled by about 5 to 20 K before a significant number of nuclei appear which possess the critical dimensions required for stability and further growth. If a nucleating agent is added to the system, crystallization can be induced at higher temperatures. This is known as hetero-geneous nucleation and only affects the crystallization rate, not the spherulitic growth rate, at a given temperature.

10.4 Melting

The melting of a perfectly crystalline substance is an equilibrium process characterized by a marked volume change and a well-defined melting temperature. Polymers are never perfectly crystalline, but contain disordered regions and crystallites of varying size. The process is normally incomplete because crystallization takes place when the polymer is a viscous liquid. In this state, the chains are highly entangled, and as sufficient time must be allowed for the chains to diffuse into the three-dimensional order required for crystallite formation, the crystalline perfection of the sample is affected by the thermal history. Thus, rapid cooling from the melt usually prevents the development of significant crystallinity. The result is that melting takes place over a range of temperatures, and this range is a useful indication of sample crystallinity.

Effect of crystallite size on melting. The range of temperature, which covers the melting of a polymer, is indicative of the size and perfection of the crystallites in the sample. This is illustrated in a study of the melting of natural rubber samples, which has shown that the melting range is a function of the temperature of crystallization. At low crystallization temperatures the nucleation density in the rubber melt is high, segmental diffusion rates are low, and small imperfect crystalline regions are formed. Thus broad melting ranges are measured for samples crystallized at these lower temperatures, and these become narrower as the crystallization temperature increases.

This suggests that careful annealing at the appropriate temperature could produce samples with a high degree of crystallinity. These samples might then exhibit almost perfect first order phase changes at the melting temperature. A close approximation to these conditions has been attained by Mandelkern, who annealed a linear polyethylene for 40 days. The improvement in the crystalline organization is obvious from examination of the resulting fusion curves in figure 10.2, where the variation of specific volume with temperature for this sample is compared with that for a branched polyethylene of low crystallinity. The effect of branching is to decrease the percentage crystallinity, broaden the melting range, and reduce the average melting temperature. The points A and B in the diagram represent the temperatures at which the largest crystallites disappear and are regarded as the respective melting temperatures T_m for the samples.

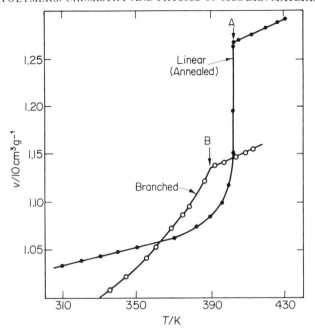

FIGURE 10.2. Specific volume v plotted against temperature T for a sample of linear polyethylene annealed for 40 days and a branched sample. Points A and B are the respective melting temperatures. (From data by Mandelkern.)

The effect of crystal size on T_m is shown more clearly in figure 10.3. The small crystals melt about 30 K lower than the large ones due to the greater contribution from the interfacial free energy in the smaller crystallites, *i.e.* there is an excess of free energy associated with the disordered chains emerging from the ends of ordered crystallites and this is relatively greater for the small crystallites, resulting in lower melting temperatures.

10.5 Thermodynamic parameters

Even with carefully annealed specimens, it is thought that the equilibrium melting temperature of the completely crystalline polymer T_m° is never actually attained. The temperature T_m° is related to the change in enthalpy ΔH_u and the entropy change ΔS_u, for the first order melting transition of pure crystalline polymer to pure amorphous melt, by

$$T_m^\circ = \Delta H_u/\Delta S_u. \tag{10.1}$$

The enthalpy change can be estimated by adding varying quantities of a diluent to the polymer, which serves to depress the observed melting temperature, and measuring T_m for each polymer + diluent mixture. The results are then plotted

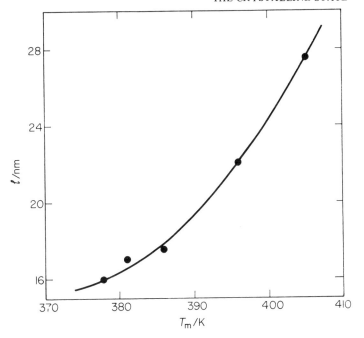

FIGURE 10.3. Dependence of T_m on the length l of the crystallite.

according to the Flory equation

$$(1/\phi_1)(1/T_m - 1/T_m^\circ) = (R/\Delta H_u)(V_u/V_1)(1 - BV_1\phi_1/RT_m), \qquad (10.2)$$

where (V_u/V_1) is the ratio of the molar volume of the repeating unit in the chain to that of the diluent, and ϕ_1 is the volume fraction of the diluent. The factor (BV_1/RT_m) is equivalent to the Flory interaction parameter χ_1, indicating that equation (10.2) is dependent on the polymer-diluent interaction. For practical purposes T_m° is taken to be the melting temperature of the undiluted polymer irrespective of the crystalline content. Typical values obtained in this way are shown in table 10.1.

In many cases the entropy change is the most important influence on the magnitude of the melting temperature of a polymer. A large part of this entropy is due to the additional freedom which allows the chain conformational changes to occur in the melt, after the restrictions of the crystalline lattice. In the crystalline phase the chain bonds are in their lowest energy state. If the energy difference between the rotational states $\Delta\epsilon$ is low, the population of the higher energy states will increase in the melt and considerable flexing of the chain is achieved. The contribution to ΔS_u is then high. When $\Delta\epsilon$ is large, the tendency to populate the high energy states is not too great, consequently the chain is less flexible and ΔS_u is lower. Two polymers which exist in the all *trans* state in

TABLE 10.1. Thermodynamic parameters derived from melting; the quantities refer to unit amount of basic unit shown

Polymer	$\dfrac{T_m}{K}$	$\dfrac{\Delta H_u}{J\ mol^{-1}}$	$\dfrac{\Delta S_u}{J\ K^{-1}\ mol^{-1}}$	Basic Unit
polyethylene	410	3970	9.70	$\left(\!-CH_2-\right)$
poly(tetrafluoroethylene)	645	2860	4.76	$\left(\!-CF_2-\right)$
cis-1,4-polyisoprene	301	4400	14.60	$\left(\!-CH_2-\underset{\underset{CH_3}{\vert}}{C}=CH-CH_2-\right)$
trans-1,4-polyisoprene	347	12700	36.90	
polypropylene	447	10880	24.40	$\left(\!-CH_2-\underset{\underset{CH_3}{\vert}}{CH}-\right)$
poly(decamethylene terephthalate)	411	46000	114.00	$\left[-(CH_2)_{10}-O-\underset{\underset{O}{\parallel}}{C}-\langle\!\!\!\bigcirc\!\!\!\rangle-\underset{\underset{O}{\parallel}}{C}-O-\right]$

the crystal are polyethylene and poly(tetrafluoroethylene). For polyethylene $\Delta\epsilon$ is about 3.0 kJ mol^{-1}, but it is as high as 18.0 kJ mol^{-1} for poly(tetrafluoro-ethylene). Hence the polyethylene chain is much more flexible in the melt and gains considerably more entropy on melting, so that T_m is correspondingly lower.

10.6 Crystalline arrangement of polymers

The formation of stable crystalline regions in a polymer requires that, (i) an economical close packed arrangement of the chains can be achieved in three dimensions, and that (ii) a favourable change in internal energy is obtained during this process. This imposes restrictions on the type of chain which can be crystallized with ease and, as mentioned earlier, one would expect symmetrical linear chains such as polyesters, polyamides, and polyethylene to crystallize most readily.

FACTORS AFFECTING CRYSTALLINITY AND T_m
These can be dealt with under the general headings, symmetry, intermolecular bonding, tacticity, branching and molar mass.

Symmetry. The symmetry of the chain shape influences both T_m and the ability to form crystallites. Polyethylene and poly(tetrafluoroethylene) are both sufficiently symmetrical to be considered as smooth stiff cylindrical rods. In the crystal these rods tend to roll over each other and change position when thermally agitated. This motion within the crystal lattice, called *premelting*, increases the entropy of the crystal and effectively stabilizes it. Consequently, more thermal energy is required before the crystal becomes unstable, and T_m is raised. Flat or irregularly shaped polymers, with bends and bumps in the chain, cannot move in this way without disrupting the crystal lattice, and so have lower T_m values. This is only one aspect.

For crystallite formation in a polymer, easy close-packing of the chains in a regular three-dimensional fashion is required. Again linear symmetrical molecules are best. Polyethylene, poly(tetrafluoroethylene) and other chains with more complex backbones containing $\{$ O $\}$, $\{$ COO $\}$, and $\{$ CONH $\}$ groups all possess a suitable symmetry for crystallite formation and usually assume extended zig-zag conformations when aligned in the lattice.

Chains containing irregular units, which detract from the linear geometry, reduce the ability of a polymer to crystallize. Thus *cis*-double bonds (I), *o*- and *m*-phenylene groups (II), or *cis*-oriented puckered rings (III), all encourage

I   II   III

bending and twisting in the chains and make regular close-packing very difficult. If, however, the phenylene rings are *para*-oriented, the chains retain their axial symmetry and can crystallize more readily. Similarly, incorporation of a *trans-*

double bond maintains the chain symmetry. This is highlighted when comparing the amorphous elastomeric *cis*-polyisoprene with the highly crystalline *trans*-polyisoprene which has no virtue as an elastomer, or *cis*-poly(1,3-butadiene) T_m = 262 K, with *trans*-poly(1,3-butadiene), T_m = 421 K.

Intermolecular bonding. In polyethylene crystallites, the close packing achieved by the chains allows the van der Waals forces to act co-operatively and provide additional stability to the crystallite. Any interaction between chains in the crystal lattice will help to hold the structure together more firmly and raise the melting temperature. Polymers containing polar groups, *e.g.* Cl, CN, or OH, can be held rigid, and aligned, in a polymer matrix by the strong dipole-dipole interactions between the substituents, but the effect is most obvious in the symmetrical polyamides. These polymers can form intermolecular hydrogen bonds which greatly enhance crystallite stability. This is illustrated in figure 10.4 for nylon-6,6, where the extended zig-zag conformation is ideally suited to allow regular intermolecular hydrogen bonding. The increased stability is reflected in T_m, which for nylon-6,6 is 540 K compared with 410 K for polyethylene

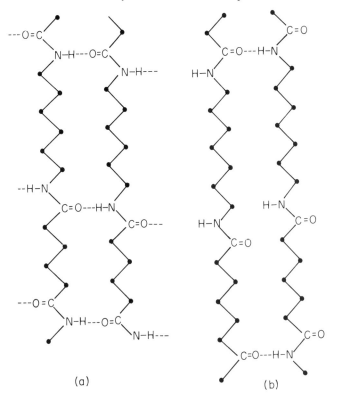

(a)   (b)

FIGURE 10.4. Extended zig-zag structures for (a) nylon-6,6 and (b) nylon-7,7 showing the allowed hydrogen bonding.

The structures of related polyamides do not always lead to this neat arrangement of intermolecular bond formation; for example the geometry of an extended nylon-7,7 chain allows the formation of only every second possible hydrogen bond when the chains are aligned and fully extended. However, the process is so favourable energetically, that sufficient deformation of the chain takes place to enable formation of all possible hydrogen bonds. The added stability that this imparts to the crystallite far outweighs the limited loss of energy caused by chain flexing.

Secondary bonds can therefore lead to a stimulation of the crystallization process in the appropriate polymers.

Tacticity. Chain symmetry and flexibility both affect the crystallinity of a polymer sample. If a chain possesses large pendant groups, these will increase the rigidity but also increase the difficulty of close packing to form a crystalline array. This latter problem can be overcome if the groups are arranged in a regular fashion along the chain. Isotactic polymers tend to form helices to accommodate the substituents in the most stable steric positions; these helices are regular forms capable of regular alignment. Thus atactic polystyrene is amorphous but isotactic polystyrene is semi-crystalline (T_m = 513 K).

Syndiotactic polymers are also sufficiently regular to crystallize, but not necessarily as a helix, rather in glide planes.

Branching in the side group tends to stiffen the chain and raise T_m, as shown in the series poly(but-1-ene), T_m = 399 K; poly(3-methyl but-1-ene), T_m = 418 K; poly(3,3'-dimethyl but-1-ene), T_m > 593 K. If the side group is flexible and non-polar, T_m is lowered.

Branching and molar mass. If the chain is substantially branched, the packing efficiency deteriorates and the crystalline content is lowered. Polyethylene provides a good example of this (figure 10.2) where extensive branching lowers the density and T_m of the polymer.

Molar mass can also alter T_m. Chain ends are relatively free to move and if the number of chain ends is increased by reducing the molar mass, then T_m is lowered because of the decrease in energy required to stimulate chain motion and melting. For example, polypropylene, with M = 2000 g mol^{-1}, has T_m = 387 K, whereas a sample with M = 30000 g mol^{-1}, has T_m = 443 K.

10.7 Morphology and kinetics
Having once established that certain polymeric materials are capable of crystallizing, fundamental studies are directed along two main channels of interest centred on (a) the mode and kinetics of crystallization, and (b) the morphology of the sample on completion of the process.

Although the morphology depends largely on the crystallizing conditions, we shall consider the macro- and microscopic structure first before dealing with the kinetics of formation.

10.8 Morphology

A number of distinct morphological units have been identified during the crystallization of polymers from the melt, which have helped to clarify the mechanism. We shall now discuss the ordered forms which have been identified.

Crystallites. In an X-ray pattern produced by a semicrystalline polymer, the discrete maxima observed arise from the scattering by small regions of three-dimensional order, which are called crystallites. They are formed in the melt by diffusion of molecules, or sections of molecules, into close packed ordered arrays; these then crystallize. The sizes of these crystallites are small relative to the length of a fully extended polymer chain, but they are also found to be independent of the molar mass and rarely exceed 1 to 100 nm. As a result, various portions of one chain may become incorporated in more than one crystallite during growth, thereby imposing a strain on the polymer which retards the process of crystallite formation. This will also introduce imperfections in the crystallites which continue growing until the strains imposed by the surrounding crystallites eventually stop further enlargement. Thus a matrix of ordered regions with disordered interfacial areas is formed, but, unlike materials with small molar masses, the ordered and disordered regions are not discrete entities and cannot be separated by differential solution techniques unless the solvent causes selective degradation of the primary bonds in the amorphous regions.

Crystallites of cellulose have been isolated from wood pulp in this way by treatment with acid to hydrolyse and remove the amorphous regions. Typical dimensions of the remaining crystallites were 46 nm long by 7.3 nm wide corresponding to bundles of about 100 to 150 chains in each crystallite.

The first attempts to explain the crystalline structure of a polymer sample produced a model called the fringe-micelle structure. The chain was envisaged as meandering throughout the system, entering and leaving several ordered regions along its length. The whole structure was thus made up of crystalline regions imbedded randomly in a continuous amorphous matrix. This model has now been virtually discarded in the light of more recent research which has revealed features incompatible with this picture.

Single crystals. When a polymer is crystallized from the melt, imperfect poly-crystalline aggregations are formed in association with a substantial amorphous content. This is a consequence of chain entanglement and the high viscosity of the melt combining to hinder the diffusion of chains into the ordered arrays necessary for crystallite formation.

If these restrictions to free movement are reduced and a polymer is allowed to crystallize from a dilute solution, it is possible to obtain well-defined single crystals. By working with solutions in which the amount of polymer is considerably less than 0.1 per cent the chance of a chain being incorporated in more than one crystal is greatly reduced, thereby increasing the possibility of isolated single crystals being formed.

These crystals are usually very small, but they have been detected for a range of polymers including polyesters, polyamides, polyethylene, cellulose acetate, and poly(4-methyl pentene-1). Although small, these single crystals can be studied using an electron microscope. This reveals that they are made up of thin lamellae, often lozenge shaped, sometimes oval, about 10 to 20 nm thick, depending on the temperature of crystallization. The most surprising feature of these lamellae is that while the molecular chains may be as long as 1000 nm, the direction of the chain axis is across the thickness of the platelet. This means that the chain must be folded many times like a concertina to be accommodated in the crystal.

For a polymer such as polyethylene, the fold in the chain is completed using only 3 or 4 monomer units with bonds in the *gauche* conformation. The extended portions in between have about 40 monomers units all in the *trans* conformation.

The crystals, thus formed, have a hollow pyramid shape, because of the requirement that the chain folding must involve a staggering of the chains if the most efficient packing is to be achieved. There is also a remarkable constancy of lamellar thickness, but this increases as the temperature increases. While opinions vary between kinetic and thermodynamic reasons for this constancy of fold distance, it is suggested that the fold structure allows the maximum amount of crystallization of the molecule at a length which produces a free energy minimum in the crystal. One suggestion is that the folding maintains the appropriate kinetic unit of the chain at any given temperature; as this would be expected to lengthen with increasing temperature, it would account for the observed thickening of the lamellae.

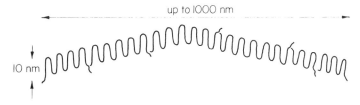

FIGURE 10.5. Idealized cross-section through a single crystal, showing folded chain structure and hollow pyramidal shape.

Measurements of the enthalpy of fusion of single crystals have shown that a significant amorphous content is still associated with each crystal. This must be on the surface and it has been suggested that it arises from the disordered folding planes produced by the emergence and re-entry of a chain in the crystal. Three models to describe this have been suggested:

(1) *Regular*, with adjacent re-entry folds;
(2) *Irregular,* in which the folding, though still adjacent, is made up of unequal lengths of chain;
(3) *Switchboard,* where the entry is random.

These are illustrated in figure 10.6, and current thinking favours the regular model for crystals grown from dilute solution.

The inadequacy of the fringe-micelle theory now becomes more obvious and modern concepts centre on the models described above. A judicious compromise of the more extreme proposition that a crystalline polymer is a single phase with defects leads to a reasonable view that this only applies to very highly crystalline polymers. The less crystalline samples are considered to be two phase systems, one ordered, one disordered, in which the loops and folds on the crystal surface, together with the defects, contribute to an additional amorphous phase.

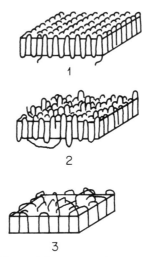

FIGURE 10.6. Three possible models, suggested to represent lamellar crystal theories: 1, regular, structure with adjacent re-entry folds; 2, irregular, with adjacent re-entry folds of varying lengths; 3, switchboard with non-adjacent fold-entry.

Hedrites. If the concentration of the polymer solution is increased a crystalline polyhedral structure emerges composed of lamellae joined together along a common plane. These have also been detected growing from a melt which suggests that lamellar growth can take place in the melt and may be a sub-unit of the spherulite.

Spherulites. Examination of thin sections of semicrystalline polymers reveals that the crystallites themselves are not arranged randomly, but form regular birefringent structures with circular symmetry. These structures, which exhibit a characteristic Maltese cross optical extinction pattern, are called spherulites. While spherulites are characteristic of crystalline polymers, they have also been observed to form in low molar mass compounds which are crystallized from highly viscous media.

Each spherulite grows radially from a nucleus formed either by the density fluctuations which result in the initial chain ordering process or from an impurity in the system. As the structure is not a single crystal, the sizes found vary from somewhat greater than a crystallite to diameters of a few millimetres. The number, size, and fine structure depend on the temperature of crystallization, which determines the critical size of the nucleating centre. This means that large fibrous structures form near T_m, whereas greater numbers of small spherulites grow at lower temperatures. When the nucleation density is high, the spherical symmetry tends to be lost as the spherulite edges impinge on their neighbours to form a mass such as shown in figure 10.7.

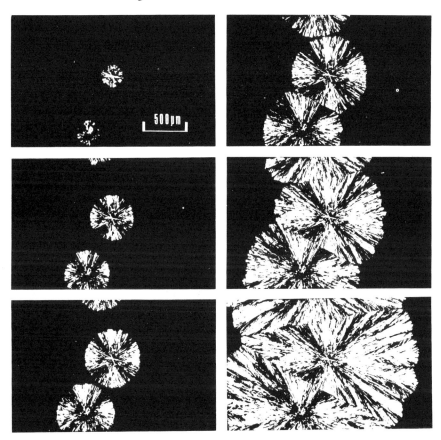

FIGURE 10.7. Sequence of photographs taken, under polarized light, over a period of about 1 min, showing the growth of poly(ethylene oxide) spherulites from the melt. From top left to bottom right; initially discrete spherulites with spherical symmetry are observed but the growing fronts eventually impinge on one another to form an irregular matrix. (Photographer: R. B. Stewart.)

A study of the fine structure of a spherulite shows that it is built up of fibrous sub-units, growth takes place by the formation of fibrils which spread outwards from the nucleus in bundles, into the surrounding amorphous phase. As this fibrillar growth advances, branching takes place, and at some intermediate stage in the development, the spherulite often resembles a sheaf of grain, figure 10.8. This forms as the fibrils fan out and begin to create the spherical outline. Although the fibrils are arranged radially, the molecular chains lie at right angles to the fibril axis. This has led to the suggestion that the fine structure is created from a series of lamellar crystals winding helically along the spherulite radius.

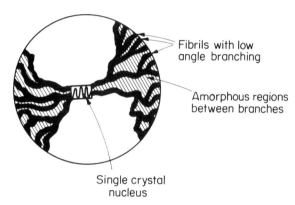

Fibrils with low
angle branching

Amorphous regions
between branches

Single crystal
nucleus

FIGURE 10.8. Schematic diagram of a spherulite based on the structure proposed by Sharples, showing crystalline fibrils emanating from a single crystal nucleus, interspersed with amorphous regions.

Spherulites are classified as positive when the refractive index of the polymer chain is greater across the chain than along the axis, and negative when the greater refractive index is in the axial direction. They also show various other features such as zig-zag patterns, concentric rings, and dendritic structures.

Sharples has proposed a picture of the main morphological features produced during the crystallization from the melt of a polymer, which are shown schematic-ally in figure 10.8, as a spherulite which he suggests is the major unit. Growth proceeds from a small crystal nucleus which develops into a fibril. Low branching and twisting then produces bundles of diverging and spreading fibrils which eventually fill out into the characteristic spherical structure. In between the branches of the fibrils are amorphous areas and these, along with the crystal defects, make up the disordered content of the semi-crystalline polymer.

10.9 Kinetics of crystallization

The crystalline content of a polymer has a profound effect on its properties and it is important to know how the rate of crystallization will vary with the temperature, especially during the processing and manufacturing of polymeric articles. The chemical structure of the polymer is also an important feature in the crystallization; for example, polyethylene crystallizes readily and cannot be quenched rapidly enough to give a largely amorphous sample whereas this is readily accomplished for isotactic polystyrene. However, this aspect will be discussed more fully later.

Isothermal crystallization. Two main factors influence the rate of crystallization at any given temperature: (i) the rate of nucleation; and (ii) the subsequent rate of growth of these nuclei to macroscopic dimensions.

The kinetic treatment of crystallization from the melt is based on the radial growth of a front through space and can be likened to someone scattering a handful of gravel onto the surface of a pond. Each stone is a nucleus which, when it strikes the surface, generates expanding circles (similar to spherulites in two dimensions). These grow unimpeded for a while but the leading edges eventually collide with others and growth rates are altered. When a similar picture is adopted for the crystallization of a polymer certain basic assumptions are made first.

The formation of ordered growth centres by the alignment of chains from the melt is called *spontaneous nucleation*. When the temperature of crystallization is close to the melting temperature, nucleation is sporadic and only a few large spherulites will grow. At lower temperatures, nucleation is rapid and a large number of small spherulites are formed. The growth of the spherulites may occur in one, two, or three dimensions and the rate of radial growth is taken to be linear at any temperature. Finally the density ρ_c of the crystalline phase is considered to be uniform throughout but different from that of the melt ρ_L. A kinetic treatment has been developed taking account of these points.

The Avrami equation. The kinetic approach relies on the establishment of a relation between the density of the crystalline and melt phases and the time. This provides a measure of the overall crystallization rate. It is assumed that the spherulites grow from nuclei whose relative positions in the melt remain unaltered, and the analysis allows for the eventual impingement of the growing discs on one another. The final relation describing the process is known as the Avrami equation expressed as

$$w_L/w_0 = \exp(-kt^n), \tag{10.3}$$

where k is the rate constant, w_0 and w_L are the masses of the melt at zero time and that left after time t. The exponent n is the Avrami exponent and is an integer which can provide information on the geometric form of the growth.

TABLE 10.2. Relation between the Avrami exponent and the morphological unit formed for sporadic nucleation

Growth unit	Nucleation	Avrami exponent n
Fibril	sporadic	2
Disc	sporadic	3
Spherulite	sporadic	4
Sheaf	sporadic	6

Sporadic nucleation is assumed to be a first-order mechanism and if we consider that a two-dimensional disc is formed, then $n = 2 + 1 = 3$. Rapid nucleation, is a zeroth-order process, in which the growth centres are formed at the same time and for each growth unit listed in the table 10.2 the corresponding values of the exponent would be $(n - 1)$. Thus the Avrami exponent is the sum of the order of the rate process and the number of dimensions the morphological unit possesses.

Dilatometry. As crystallization involves the close packing of chains in regular three-dimensional structures, the economical use of space is accompanied by an increase in density. Thus the rate of crystallization can be followed by recording the density changes which are readily detected in a dilatometer. This is achieved by placing the polymer in a dilatometer with a confining liquid, such as mercury, so that any volume change can be recorded as a movement of the liquid meniscus in a capillary. A typical design is shown in figure 10.9.

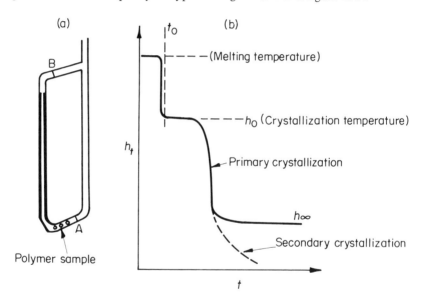

FIGURE 10.9. (a) Typical dilatometer for following crystallization kinetics and (b) general shape of a plot of dilatometer height h_t as a function of time t obtained in crystallization experiments.

The polymer is introduced into the dilatometer between the point A and the capillary. The apparatus is then pumped out and sealed under vacuum at the point A. Sufficient mercury is then added to enclose the polymer and extend into the capillary, after which the tube is sealed at B, and placed in a thermostat at a temperature somewhat higher than the melting temperature of the polymer. When the sample is completely molten the dilatometer is transferred to a second thermostat set at the temperature selected for crystallization to take place and allowed to equilibrate. The initial period of temperature adjustment to the second temperature may make the initial height h_0 rather difficult to locate, but usually a plot such as shown in figure 10.9(b) is recorded. If secondary crystallization takes place the final portion of the curve may tail away making h_∞ more difficulty to measure.

The mass fraction of the uncrystallized polymer ($w_L w_0$) can be related to the volume changes and to the heights measured in the dilatometer by

$$w_L/w_0 = (V_t - V_\infty)/(V_0 - V_\infty) = (h_t - h_\infty)/(h_0 - h_\infty) = \exp(-kt^n), \qquad (10.4)$$

where h_t, h_0, and h_∞ are the heights at time t, the beginning, and the end of the process respectively, with V_t, V_0 and V_∞ the corresponding volumes. The slope of a plot of $-\ln\{(h_t - h_\infty)/(h_0 - h_\infty)\}$ against t allows evaluation of the Avrami exponent n while k can be calculated from the intercept.

Deviations from Avrami equation. The Avrami equation can describe some but not all systems investigated. The crystallization isotherms of poly(ethylene terephthalate) can be fitted by equation (10.3) using $n = 4$ above 473 K and $n = 2$ at 383 K. The equation should be used with caution, however, as non-integer values have been reported and the geometric shape of the morphological unit is not always that predicted by the value of n calculated from the experimental data.

Secondary crystallization. Deviations from the Avrami treatment may also be observed towards the end of the crystallization process and values of h_∞ are often difficult to determine accurately, as shown in the curve derived from dilatometric data. The tailing of the curve is a result of a secondary crystallization process which is a slower reorganization of the crystalline regions to produce more perfectly formed crystallites.

General reading

P. Geil, *Polymer Single Crystals.* Interscience Publishers Inc. (1963).
M. Gordon, *High Polymers.* Iliffe Books (1963).
L. Mandelkern, *Crystallization of Polymers.* McGraw-Hill (1964).
P. Meares, *Polymers: Structure and Bulk Properties*, Chapters 4 and 5. Van Nostrand (1965).
A. Sharples, *Introduction to Polymer Crystallization.* Edward Arnold (1966).
D. A. Smith, *Addition Polymers*, Chapter 8. Butterworths (1968).
A. V. Tobolsky and H. Mark, *Polymer Science and Materials*, Chapter 8. Wiley-Interscience (1971).

References

1. L. Manderlkern, *Rubber Chem. Technol.*, 32, 1392 (1949).
2. L. Marker, R. Early and S. L. Aggarwal, *J. Polym. Sci.*, 38, 369 (1959).

CHAPTER 11

The Amorphous State

11.1 Molecular motion

A linear polymer chain can be treated as a "one-dimensional co-operative system" in which the rotation of a chain segment is restricted or aided by the neighbouring segments. For long chains, co-operative motion cannot be expected to extend along the entire length, and the polymer tends to act as if it were composed of a series of interconnected, but independent, kinetic units. Any significant movement of such a chain is generated by rotation about the single bonds connecting the atoms in the chain, and depends on the ease of interchange of any element from one rotational state to another. The height of the potential energy barrier ΔE (*cf.* figure 1.3) will determine the rapidity of conformational change at any temperature, and when the temperature of the polymer increases, the additional thermal energy allows ΔE to be overcome more often. This encourages increasing molecular motion until eventually the polymer behaves like a viscous liquid (assuming that no thermal degradation takes place).

In the amorphous state the distribution of polymer chains in the matrix is completely random, with none of the strictures imposed by the ordering encountered in the crystallites of partially crystalline polymers. This allows the onset of molecular motion in amorphous polymers to take place at temperatures below the melting temperature of such crystallites. Consequently, as the molecular motion in an amorphous polymer increases, the sample passes from a glass, through a rubber-like state, until finally it becomes molten. These transitions lead to changes in the physical properties and material application of a polymer, and it is important to examine physical changes wrought in an amorphous polymer as a result of variations in the molecular motion.

11.2 The five regions of viscoelastic behaviour

The physical nature of an amorphous polymer is related to the extent of the molecular motion in the sample, which in turn is governed by the chain flexibility and the temperature of the system. Examination of the mechanical behaviour shows that there are five distinguishable states in which a linear amorphous polymer can exist and these are readily displayed if a parameter such as the elastic modulus is measured over a range of temperatures.

The general behaviour of a polymer can be typified by results obtained for an amorphous atactic polystyrene sample. The relaxation modulus E_r was measured at a standard time interval of 10 s and $\log_{10} E_r$ is shown as a function of temperature in figure 11.1. Five distinct regions can be identified on this curve.

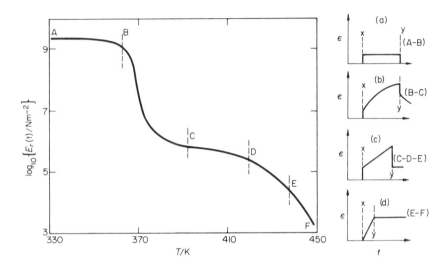

FIGURE 11.1. Five regions of viscoelasticity, illustrated using a polystyrene sample. Also shown are the strain—time curves for stress applied at x and removed at y: (a) glassy region; (b) leathery state; (c) rubbery state; and (d) viscous state.

(i) *The glassy state.* This is section A to B lying below 363 K and it is characterized by a modulus between $10^{9.5}$ and 10^9 N m^{-2}. Here co-operative molecular motion along the chain is frozen, causing the material to respond like an elastic solid to a stress, and the strain—time curve is of the form shown in figure 11.1(a).
(ii) *Leathery or retarded highly elastic state.* This is the transition region B to C where the modulus drops sharply from about 10^9 to about $10^{5.7}$ N m^{-2} over the temperature range 363 to 393 K. The glass transition temperature T_g is located in this area and the rapid change in modulus reflects the constant increase in molecular motion as the temperature rises from T_g to about $(T_g + 30$ K$)$. Just above T_g the movement of the chain segments is still rather slow, imparting what can best be described as leathery properties to the material. The strain—time curve is that shown in figure 11.1(b).

(iii) *The rubbery state.* At approximately 30 K above the glass transition the modulus curve begins to flatten out into the plateau region C to D in the modulus interval $10^{5.7}$ to $10^{5.4}$ N m^{-2} and extends up to about 420 K.

(iv) *Rubbery flow.* After the rubbery plateau the modulus again decreases from $10^{5.4}$ to $10^{4.5}$ N m^{-2} in the section D to E. The effect of applied stress to a polymer in states (iii) and (iv) is shown in figure 11.1(c) where there is instantaneous elastic response followed by a region of flow.

(v) *Viscous State.* Above a temperature of 450 K, in the section E to F, there is little evidence of any elastic recovery in the polymer and all the characteristics of a viscous liquid become evident (figure 11.1(d)). Here there is a steady decrease of the modulus from $10^{4.5}$ N m^{-2} as the temperature increases.

The overall shape of the curve shown in figure 11.1 is typical for linear amorphous polymers in general, although the temperatures quoted are specific to polystyrene and will differ for other polymers. Variations in shape are found for different molar masses and when the sample is crosslinked or partly crystalline. The value of the modulus provides a good indication of the state of the polymer and can be obtained from the curve.

11.3 The viscous region

Before considering the flow in polymer melts, the viscous behaviour of simple liquids will be examined.

The application of a force to a simple liquid of low molar mass is relieved by the flow of molecules past one another into new positions in the system. A liquid, forced to flow in this way by a shearing force σ, experiences a viscous resistance expressed by

$$\eta = \sigma(dv/dx)^{-1}, \qquad (11.1)$$

where v is the velocity of flow along a tube of radius x, so that (dv/dx) is the velocity gradient or shear rate $\dot{\gamma}$, and η is the viscosity coefficient of the liquid. A liquid is said to exhibit Newtonian flow if η is independent of $\dot{\gamma}$ but substances which show deviations from this flow pattern, with either decreasing or increasing $(\sigma/\dot{\gamma})$ ratios, are termed non-Newtonian. (See figure 11.2.) Most polymers fall into this latter category, with η decreasing as the shear rate increases.

The temperature dependence of η can normally be expressed in the form

$$\eta = A \exp(\Delta E_D/RT), \qquad (11.2)$$

where A is a constant and ΔE_D represents the activation energy required to create a hole big enough for a molecule to translate or "jump" into during flow. In liquids with larger or irregularly shaped molecules, the deformation is slower as the molecules restrict the easy translation of one past the other. This results in a high value of η.

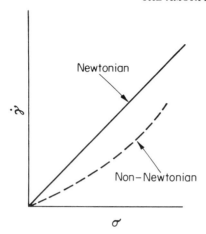

FIGURE 11.2. Newtonian and non-Newtonian flow curves.

11.4 Kinetic units in polymer chains

Resistance to flow in polymer systems is even greater, because now the molecules are covalently bonded into long chains which are coiled and entangled and translational motion must, of necessity, be a co-operative process. It would be unreasonable to expect easy co-operative motion along the entire polymer chain, but as there is normally some degree of flexibility in the chain, local segmental motion can take place more readily. The polymer can then be considered as a series of kinetic units; each of these moves in an independent manner and involves the co-operative movement of a number of consecutive chain atoms. *Crankshaft Motion.* If we now consider an arbitrary kinetic unit which involves the movement of six atoms by rotation about two chain bonds, the movement can be visualized as shown diagrammatically in figure 11.3. The amorphous or

FIGURE 11.3. Crankshaft motion in a polymer chain.

molten polymer is a conglomeration of badly packed interlacing chains and the extra empty space caused by this random molecular arrangement is called the *free volume* which essentially consists of all the holes in the matrix. When sufficient thermal energy is present in the system the vibrations can cause a segment to jump into a hole by co-operative bond rotation and a series of such jumps will enable the complete polymer chain eventually to change its position. Heating will cause a polymer sample to expand thereby creating more room for movement of each kinetic unit and the application of a stress in a particular direction will encourage flow by segmental motion in the direction of the stress. The segmental transposition involving six carbon atoms is called crankshaft motion and is believed to require an activation energy of about 25 kJ mol⁻¹.

11.5 Effect of chain length

Although it is thought that translation of a polymer chain proceeds by means of a series of segmental jumps involving short kinetic units, which may each consist of between 15 and 30 chain atoms, the complete movement of a chain cannot remain unaffected by the surrounding chains. As stated previously, considerable entanglement exists in the melt and any motion will be retarded by other chains.

According to Bueche the polymer molecule may drag along several others during flow and the energy dissipation is then a combination of the friction between the chain plus those which are entangled and the neighbouring chains as they slip past each other. It would seem reasonable to assume from this, that the length of the chains in the sample must play a significant role in determining the resistance to flow and the effect of chain length on $\log \eta$, measured at low shear rates to ensure Newtonian flow, is illustrated in figure 11.4. The plot comprises two linear portions meeting at a critical chain length Z_c. Above Z_c the relation describing the flow behaviour is

$$\log \eta = 3.4 \log Z + \log K_2, \tag{11.3}$$

and η is proportional to the 3.4 power of Z. Below Z_c, η is directly proportional to Z and the expression becomes

$$\log \eta = \log Z + \log K_1, \tag{11.4}$$

where K_1 and K_2 are temperature dependent constants.

The critical chain length Z_c is interpreted as representing the dividing point between chains which are too short to provide a significant contribution to η from entanglement effects and those large enough to cause retardation of flow by intertwining with their neighbours. If Z is defined as the number of atoms in the backbone chain of a polymer then typical values for Z_c are 610 for polyisobutylene, 730 for polystyrene, and 208 for poly(methyl methacrylate). In general Z_c is lower for polar polymers than for non-polar polymers.

11.6 Temperature dependence of η

When a polymer is transformed into a melt without degradation and is stable at even higher temperatures, η is observed to decrease rapidly as the temperature increases. If it is still stable at temperatures in excess of 100 K above T_g, the temperature dependence has an exponential form

$$\eta = B \exp(\Delta H/RT), \tag{11.5}$$

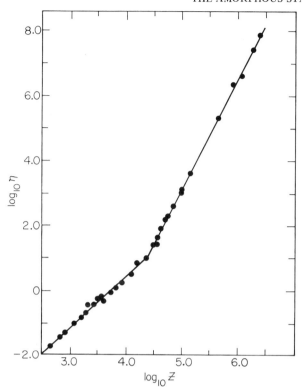

FIGURE 11.4. Dependence of melt viscosity on chain length $\mathcal{Z}$, for polyisobutylene fractions measured at low shear rates and at 490 K. (Data by Fox and Flory, 1951.)

where according to the Eyring rate theory ΔH is the activation enthalpy of viscous flow and is a more representative parameter than the energy. Values of ΔH vary slowly over a range from 20 to 120 kJ mol^{-1}. When the temperature is lowered towards T_g, ΔH changes dramatically and a simple equation such as (11.5) is no longer valid. The increase in ΔH, observed with temperature lowering, can be equated with a rapid loss of free volume as T_g is approached. Hence ΔH becomes dependent on the availability of a suitable hole for a segment to move into, rather than being representative of the potential energy barrier to rotation. This approach suggests that the jump frequency decreases when there is an increasing co-operative motion among the chains needed to produce holes.

The WLF equation. An expression is now required which will predict the behaviour of highly viscous polymers in the range T_g to $(T_g + 100\ \text{K})$. The one proposed by Williams, Landel, and Ferry has proved most successful, and can be expressed as

$$\log_{10}(\eta_T/\eta_{T_0}) = \frac{[-a_1(T - T_0)]}{[a_2 + T - T_0]}\left(\frac{T\rho}{T_0\rho_0}\right). \tag{11.6}$$

This is usually referred to as the WLF equation in which T_0 is a reference temperature, often chosen as the glass temperature T_g. When this is the case the constants assume the values $a_1 = 17.44$ and $a_2 = 51.6$ K. The final factor is a small correction factor to compensate for a changing density ρ, with temperature. The ratio log (η_T/η_{T_g}) is sometimes called the log "shift" factor and given as log a_T.

11.7 Rubbery state

With a decrease in temperature, the flow of a polymer melt becomes increasingly sluggish as the chain motion becomes too slow to effect complete untangling of the polymer coils. The viscosity increases rapidly to a value of about 10^{12} Pa s as T_g is approached, but on passing from the melt to the glass a region of rubbery flow and elasticity is traversed. In this state the polymer exhibits several unique properties which are dealt with in chapter 13 and only a brief description of the chain behaviour in this region is given here.

Long range elasticity. The rubber-like region, which lies above T_g, appears when the rotation about the segment links is free enough to enable the chains to assume any of the immense number of equi-energetic conformations available, without significant chain untangling taking place. The majority of these shapes will be compact coils because the possibility of their occurrence is much greater than for the more extended forms.

When a polymer, which is not too crystalline and has a reasonably high molar mass ($> 20\,000$ g mol^{-1}), is in this elastic state it will elongate quite readily in the direction of an applied stress, *e.g.* natural rubber will stretch easily when pulled. If the stress is applied for a short time, then removed, the sample snaps back to its original length suggesting that some "memory" of its initial unstretched condition is retained. The ability of an elastomer to regain its former size, when extensions of up to 400 per cent have been experienced, is associated with the long chain character of the material. This retractive action of linear uncrosslinked polymers can be observed if the time interval between extension and release is short, but if the stress is maintained for some time, then a relaxation process takes place allowing the tension to decay eventually to zero.

This can be explained quite simply. The molecules are initially in highly coiled shapes but application of a force causes rotation about the chain bonds resulting in an elongation of the molecules in the direction of the stress. This produces a distribution of chain conformations which differs significantly from the most probable distribution, and as this is an unstable state the chains will rapidly recoil when the stress is released in an attempt to regain their original shape distribution. For short periods of stress in an amorphous elastomer, the entanglement and intertwining of chains with their neighbours acts as a physical restraint to excessive chain movement and the elastomer regains its original length when the stress is removed. If however, the stress is maintained for sufficiently long, there is a general tendency for chains to unravel and slip past one another into new positions where the segments can relax and regain a stable coiled form. The resultant flow relieves the tension and produces the observed stress decay. The process is shown schematically in figure 11.5. When the molar mass is too low to produce sufficient entanglement, the material will flow more readily and behave like a viscous liquid. Similarly, as the temperature increases further and further above the glass transition, the enhanced segmental movement facilitates stress decay because of the greater ease of chain disentanglement.

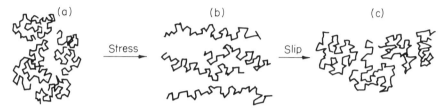

FIGURE 11.5. Schematic representations of an elastomer (a) under no stress, (b) chain alignment under an applied stress, and (c) stress relief produced by chains slipping past one another into new positions in the sample and recoiling.

11.8 Glass transition region

When the polymer is at a temperature below its glass temperature, chain motion is frozen. The polymer then behaves like a stiff spring storing all the available energy in stretching as potential energy, when work is performed on it. If sufficient thermal energy is supplied to the system to allow the chain segments to move co-operatively, a transition from the glass to the rubber-like state begins to take place. Motion is still restricted at this stage, but as the temperature increases further a larger number of chains begin to move with greater freedom. In mechanical terms the transition can be likened to the transformation of a stiff spring to weak spring. As weak springs can only store a fraction of the potential energy that a strong spring can hold, the remainder is lost as heat and if the change from a strong to a weak spring takes place over a period of time, equivalent to the observation time, then the energy loss is detected as mechanical damping. Finally when molecular motion increases to a sufficiently high level, all the chains behave like weak springs the whole time. This means that the modulus is much lower, but so too is the damping, which passes through a maximum in the vicinity of T_g. The maximum appears because the polymer is passing from the low-damping glassy state, through the high-damping transition region, to the lower-damping rubber-like state.

Treloar has described a very apt demonstration of the transition. A thin rubber rod is wound round a cylinder to create the shape of a spring and then frozen in this shape using liquid nitrogen. The cylinder (possibly of paper) is then removed leaving the rubber spring. The rubber is now in the glassy state and it acts like a stiff metal spring by regaining its shape rapidly after an extension. As the temperature is raised a gradual loss in the elastic recovery is observed after each applied stress, until a stage is reached when there is no recovery and the rubber remains in the deformed shape. With a further increase in temperature the rod straightens under its own weight and eventually regains its rubber-like elasticity at slightly higher temperatures.

THE GLASS TRANSITION TEMPERATURE, T_g

The transition from the glass to the rubber-like state is an important feature of polymer behaviour, marking as it does a region where dramatic changes in the physical properties, such as hardness and elasticity, are observed. The changes are completely reversible, however, and the transition from a glass to a rubber is a function of molecular motion, not polymer structure. In the rubber-like state or in the melt the chains are in relatively rapid motion, but as the temperature is lowered the movement becomes progressively slower until eventually the available thermal energy is insufficient to overcome the rotational energy barriers in the chain. At this temperature, which is known as the glass transition temperature T_g, the chains become locked in whichever conformation they possessed when T_g was reached. Below T_g the polymer is in the glassy state and is, in effect, a frozen liquid with a completely random structure.

Although the glass-rubber transition itself does not depend on polymer structure, the temperature at which T_g is observed depends largely on the chemical nature of the polymer chain and for most common synthetic polymers lies between 170 and 500 K. It is quite obvious that T_g is an important characteristic property of any polymer as it has an important bearing on the potential application of a polymer. Thus for a polymer with a flexible chain, such as polyisoprene, the thermal energy available at about 300 K is sufficient to cause the chain to change shape many thousands of times in a second. This polymer has T_g = 200 K. On the other hand, virtually no motion can be detected in atactic poly(methyl methacrylate) at 300 K, but at 450 K the chains are in rapid motion. In this case T_g = 378 K. This means that at 300 K polyisoprene is likely to exhibit rubber-like behaviour and be useful as an elastomer, whereas poly(methyl methacrylate) will be a glassy material. If the operating temperature was lowered to 100 K, both polymers would be glasses.

EXPERIMENTAL DEMONSTRATION OF T_g

The glass transition is not specific to long chain polymers. Any substance, which can be cooled to a sufficient degree below its melting temperature without crystallizing, will form a glass. The phenomenon can be conveniently demonstrated using glucose penta-acetate (GPA). A crystalline sample of GPA is melted, then chilled rapidly in ice-water to form a brittle amorphous mass. By working the hard material between one's fingers, the transition from glass to rubber will be felt as the sample warms up. A little perseverance, with further rubbing and pulling, will eventually result in the recrystallization of the rubbery phase, which then crumbles to a powder.

DETECTION OF T_g

The transition from a glass to a rubber-like state is accompanied by marked changes in the specific volume, the modulus, the heat capacity, the refractive index, and other physical properties of the polymer. The glass transition is not a first-order transition, in the thermodynamic sense, as no discontinuities are observed when the entropy or volume of the polymer are measured as a function of temperature. If the first derivative of the property-temperature curve is measured, a change in the vicinity of T_g is found; for this reason it is sometimes called a second-order transition. Thus while the change in a physical property can be used to locate T_g, the transition bears many of the characteristics of a relaxation process and the precise value of T_g can depend on the method used and the rate of the measurement.

Techniques for locating T_g can be divided into two categories, dynamic and static. In the static methods, changes in the temperature dependence of an intensive property, such as density or heat capacity are followed and measurements are carried out slowly, to allow the sample to equilibrate and relax at each observation temperature. In dynamic mechanical methods a rapid change in modulus is indicative of the glass transition, but now the transition region is dependent on the frequency of the applied force. If we assume that, in the transition region, the restrictions to motion still present in the sample, allow only a few segments to move in some time interval, say 10 s, then considerably fewer will have moved if the observation time is less than 10 s. This means that the location of the transition region and T_g will depend on the experimental approach used, and T_g is found to increase 5 to 7 K for every tenfold increase in the frequency of the measuring techniques. This time dependence of segmental motion corresponds to the strong-weak transformation of a hypothetical spring and results in the high damping which imparts the lifeless leathery consistency to the polymer in this region. The temperature of maximum damping is usually associated with T_g, and at low frequencies the value assigned to T_g is within a few kelvins of that obtained from the static methods. As the static methods lead to more consistent values some of these can be described.

Measurement of T_g from $V-T$ curves. One of the most frequently used methods of locating T_g is to follow the change in the volume of the polymer as a function of the temperature. The polymer sample is placed in the bulb of a dilatometer, degassed, and a confining liquid such as mercury added. If the bulb is attached to a capillary the change in polymer volume can be traced by noting the overall change in volume registered by the movement of the mercury level in the capillary. A variation of this method makes use of a density gradient column. A small sample of polymer suspended in this column provides a direct measure of the polymer density which can be measured easily as the temperature is varied.

Typical specific-volume-temperature curves are shown in figure 11.6 for poly(vinyl acetate).These consist of two linear portions whose slopes differ and closer inspection reveals that over a narrow range of temperature of between 2 and 5 K the slope changes continuously. To locate T_g, the linear portions are

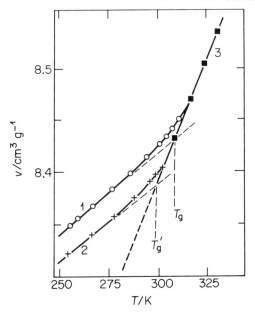

FIGURE 11.6. Specific volume v plotted against temperature for poly(vinyl acetate) measured after rapid cooling from above the T_g; 1, measured 0.02 hour after cooling; 2, measured 100 hours after cooling; T_g and T'_g are the glass transition temperatures measured for the different equilibration times. (After Kovacs, 1958.)

extrapolated and intersect at the point which is taken to be the characteristic transition temperature of the material. Each point on the curve is normally recorded after allowing the polymer time to equilibrate at the chosen temperature and as the rate of measurement affects the magnitude of T_g quite noticeably the equilibration time should be several hours at least. The effect of the measuring rate on T_g was demonstrated by Kovacs who recorded the volume of a polymer at each temperature, over a range including the transition, using two rates of cooling. If the sample was cooled rapidly (0.02 h) to each temperature the value of T_g derived from the resulting curve was some 8 K higher than that measured from results obtained using a slow cooling rate (100 h).

Refractive index measurements. The change in refractive index of the polymer with temperature has been used by several workers to establish T_g. A linear decrease in refractive index is observed as the temperature increases, and as the transition is passed, the rate of decrease becomes greater; T_g is again taken as the intersection of the linear extrapolation.
Heat capacity and other methods. The glass transition temperature can be detected calorimetrically by following the change in heat capacity with change in

temperature. The curve for atactic polypropylene is shown in figure 11.7 where the abrupt increase in c_p at about 260 K, corresponds to the glass transition.

Among other reported techniques the most useful include differential thermal analysis, dielectric loss measurements, X- and β-ray absorption, and gas permeability studies. All indicate the existence of the phenomenon which we call the glass transition.

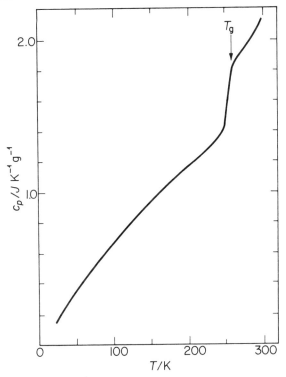

FIGURE 11.7. Specific heat capacity c_p plotted against temperature for atactic polypropylene showing the glass transiton in the region of 260 K. (O'Reilly and Karasz, 1966.)

11.9 Factors affecting T_g

We have seen that the magnitude of T_g varies over a wide temperature range for different polymers. As T_g depends largely on the amount of thermal energy required to keep the polymer chains moving, a number of factors which affect rotation about chain links, will also influence T_g. These include (1) chain flexibility, (2) molecular structure (steric effects), (3) molar mass (see section 11.11), (4) branching and crosslinking.

Chain flexibility. The flexibility of the chain is undoubtedly the most important factor influencing T_g. It is a measure of the ability of a chain to rotate about the constituent chain bonds, hence a flexible chain has a low T_g whereas a rigid chain has a high T_g.

For symmetrical polymers, the chemical nature of the chain backbone is all important. Flexibility is obtained when the chains are made up of bond sequences which are able to rotate easily, and polymers containing $+CH_2$—CH_2+, $+CH_2$—O—CH_2+, or $+Si$—O—$Si+$ links will have correspondingly low values of T_g. The value of T_g is raised markedly by inserting groups which stiffen the chain by impeding rotation, so that more thermal energy is required to set the chain in motion. The p-phenylene ring is particularly effective in this respect, but when carried to extremes, produces a highly intractable, rigid

structure, poly(p-phenylene) $+\langle\bigcirc\rangle\overline{)_n}$, with no softening point. The basic

structure can be modified by introducing flexible groups in the chain and some examples are given in table 11.1.

TABLE 11.1. Influence of bond flexibility on T_g

Polymers	Repeat unit	T_g/K
poly(dimethylsiloxane)	$\begin{array}{cc}CH_3 & CH_3 \\ \| & \| \\ +Si-O-Si+ \\ \| & \| \\ CH_3 & CH_3\end{array}$	150
polyethylene	$+CH_2-CH_2+$	180
cis-polybutadiene	$+CH_2-CH=CH-CH_2+$	188
poly(oxyethylene)	$+CH_2-CH_2-O+$	206
Poly(phenylene oxide)	$+\langle\bigcirc\rangle-O+$	356
Poly(arylene sulphone)	$+\langle\bigcirc\rangle-O-\langle\bigcirc\rangle-SO_2+$	523
poly(p-xylylene)	$+\langle\bigcirc\rangle-CH_2-CH_2+$	about 553

Steric effects. When the polymer chains are unsymmetrical, with repeat units of the type $+CH_2CHX+$, an additional restriction to rotation is imposed by steric effects. These arise when bulky pendant groups hinder the rotation about the backbone and cause T_g to increase. The effect is accentuated by increasing the size of the side group and there is some evidence of a correlation between T_g and the molar volume V_X of the pendant group. It can be seen in table 11.2, that T_g increases with increasing V_X in the progressive series, polyethylene, polypropylene, polystyrene, and poly(vinyl naphthalene). Superimposed on this group size factor are the effects of polarity and the intrinsic flexibility of the pendant group itself. An increase in the lateral forces in the bulk state will hinder molecular motion and increase T_g. Thus polar groups tend to encourage a higher T_g than non-polar groups of similar size, as seen when comparing polypropylene, poly(vinyl chloride) and polyacrylonitrile. The influence of

TABLE 11.2. Glass transition temperatures for atactic polymers of the general type $-(\ CH_2 \cdot CXY-)_n$

Polymer	T_g/K	V_X/cm^3 mol^{-1} †	Group X
Type $+CH_2CHX +_n$			
polyethylene	188	3.7	—H
polypropylene	253	25.9	—CH$_3$
poly(but-1-ene)	249	48.1	—C$_2$H$_5$
poly(pent-1-ene)	233	70.3	—C$_3$H$_7$
poly(hex-1-ene)	223	92.5	—C$_4$H$_9$
poly(4-methyl pent-1-ene)	302	92.5	—CH$_2$—CH(CH$_3$)$_2$
poly(vinyl alcohol)	358	11.1	—OH
poly(vinyl chloride)	354	22.1	—Cl
polyacrylonitrile	378	30.0	—CN
poly(vinyl acetate)	301	60.1	—O·C—CH$_3$ ‖ O
poly(methyl acrylate)	279	60.1	—C—O—CH$_3$ ‖ O
poly(ethyl acrylate)	249	82.3	—COOC$_2$H$_5$
poly(propyl acrylate)	225	104.5	—COOC$_3$H$_7$
poly(butyl acrylate)	218	126.7	—COOC$_4$H$_9$
polystyrene	373	92.3	
poly(α-vinylnaphthalene)	408	143.9	
poly(vinyl biphenyl)	418	184.0	
Type $+ CH_2C(CH_3)X +_n$		$V(X + Y)$cm^3 mol^{-1}†	
poly(methyl methacrylate)	378	86.0	
poly(ethyl methacrylate)	338	108.2	
poly(propyl methacrylate)	308	130.4	
polymethacrylonitrile	393	55.9	
poly(α-methylstyrene)	445	118.2	

† Calculated using LeBas volume equivalents. (See Glasstone "Textbook of Physical Chemistry" Macmillan, 1951, Chapter 8.)

side chain flexibility is evident on examination of the polyacrylate series from methyl through butyl, and also in the polypropylene to poly(hex-1-ene) series.

A further increase in steric hindrance is imposed by substituting an α-methyl group, which restricts rotation even further and leads to higher T_g. For the pair polystyrene-poly(α-methyl styrene), the increase in T_g is 70 K, while the difference between poly(methyl methacrylate) and poly(methyl acrylate) is 100 K.

These steric factors all affect the chain flexibility and are simply additional contributions to the main chain effects.

Configurational effects. *Cis-trans* isomerism in polydienes and tacticity variations in certain α-methyl substituted polymers alter chain flexibility and affect T_g. Some examples are shown in table 11.3. It is interesting to note that when no α-methyl group is present in a polymer, tacticity has little influence on T_g.

TABLE 11.3. The effect of microstructure on T_g

Polymer	Stereostructure	T_g/K
poly(methyl methacrylate)	isotactic	318
	atactic	378
	syndiotactic	388
polybutadiene	cis	165
	trans	255
polyisoprene	cis	200
	trans	220

Effect of crosslinks on T_g. When crosslinks are introduced into a polymer, the density of the sample is increased proportionally. As the density increases, the molecular motion in the sample is restricted and T_g rises. For a high crosslink density the transition is broad and ill-defined, but at lower values, T_g is found to increase linearly with the number of crosslinks.

11.10 Theoretical treatments

Before embarking on a rather brief description of the theoretical interpretations of the glass transition a word of caution should be given. In the foregoing sections several features of the results point to the fact that, in the vicinity of T_g, rate effects are closely associated with changes in certain thermodynamic properties. This has engendered two schools of thought on the origins of this phenomenon, together with variations on each theme. The elementary level of this text precludes detailed critical discussion of the relative merits of any particular treatment, and to avoid prejudicing the issue with personal comment the main ideas of each are outlined together with a more recent and possibly unifying approach to complete the picture.

THE FREE VOLUME THEORY

Various aspects of the free volume concept have been touched upon earlier. The free volume is the unoccupied space in a sample which arises as a result of the inefficient packing of disordered chains in the amorphous regions of a polymer. The presence of these empty volumes can be demonstrated by observing the contraction in total volume when a polystyrene glass is dissolved in benzene. This indicates that the polymer occupies less volume when efficiently surrounded by benzene molecules and that a free volume exists in the undiluted glassy state of the polymer.

Flory and Fox have suggested that a decrease in the temperature of a polymer will be accompanied by a decrease in the free volume above T_g which continues until T_g is reached. At this temperature the free volume attains a particular value which remains constant as the temperature continues to drop below T_g. The glass transition can then be pictured as a specific volume change due solely to an increase in the free volume which is shown as the shaded area in figure 11.8.

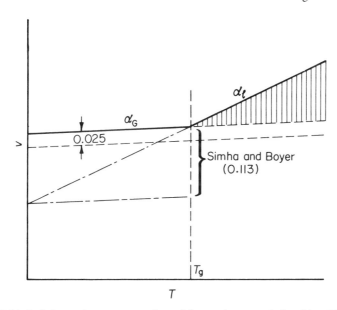

FIGURE 11.8. Schematic representation of free volume as defined by Flory and Fox (0.025) and Simha and Boyer (0.113).

This means that the specific volume of the polymer v is the sum of the unoccupied and occupied volumes v_0 and v_f so that

$$v = v_0 + v_f = v_0 + v' + \alpha_G T, \qquad (11.7)$$

where v' is the volume of the glass at 0 K and α_G is the expansivity of the glass phase, defined by

$$\alpha_G = (1/v)(dv/dT). \qquad (11.8)$$

A relation between the viscosity coefficient and the free volume has been proposed by Doolittle which can be expressed as

$$\ln \eta = \ln A' + B'(v_f/v_0).\tag{11.9}$$

If the free volume is now replaced by the free volume fraction $f = v_0(v_f + v_0)^{-1}$, this becomes

$$\ln a_T = \ln (\eta_T/\eta_{T_g}) = B'(1/f + 1/f_g),\tag{11.10}$$

where f_g is the fractional free volume at T_g. Expansion and rearrangement of equation (11.10) leads to

$$\log_{10} a_T = \left(\frac{B'}{2.303 f_g}\right) \left[\frac{(T - T_g)}{(f_g/\alpha_f) + T - T_g}\right]\tag{11.11}$$

in which α_f is the relative free volume expansion coefficient defined as $\alpha_f \approx (df/dT)$.

Equation (11.11) is similar to the WLF equation (see equation (11.6) and section 12.8) and provides some justification for the use of the semi-empirical WLF equation, which has met with considerable success when applied to a wide range of systems. In this approach the free volume fraction at T_g is assumed to have a value of $f_g = 0.025$, but the main weakness in the treatment stems from indications that this is not a universal value for all systems and although the concept is useful it is usually regarded as an approximation.

GIBBS-DI MARZIO THERMODYNAMIC THEORY

Comments on the thermodynamic theories will be restricted to the proposals of Gibbs and Di Marzio (G-D) who, while acknowledging that kinetic effects are inevitably encountered when measuring T_g, consider the fundamental transition to be a true equilibrium. The data reported by Kovacs in section 11.8 imply that the observed T_g would decrease further if a sufficiently long time for measurement was allowed. This aspect is considered in the G-D theory by defining a new transition temperature T_2 at which the configurational entropy of the system is zero. This temperature can be considered in effect to be the limiting value T_g would reach in a hypothetical experiment taking an infinitely long time. On this basis the experimentally detectable T_g is a time dependent relaxation process and the observed value is a function of the time scale of the measuring technique. The theoretical derivation is based on a lattice treatment. The configurational entropy is found by calculating the number of ways that n_x linear chains each x segments long can be placed on a diamond lattice, for which the coordination number $z = 4$, together with n_0 holes. The restrictions imposed on the placing of a chain on the lattice are embodied in the hindered rotation which is expressed as the "flex energy" $\Delta\epsilon$, and ϵ_h which is the energy of formation of a hole. The flex energy is the energy difference between the potential energy minimum of the located bond and the potential minima of the remaining $(z - 2)$ possible orientations which may be used on the lattice. Thus for polyethylene the *trans* position is considered most stable and the *gauche* positions are the flexed ones with $\Delta\epsilon$ the energy difference between the ground and flexed states. This of course varies with the nature of the polymer. The quantity ϵ_h is a measure of the cohesive energy. The configurational entropy S_{conf} is derived from the partition function describing the location of holes and polymer molecules.

As the temperature drops towards T_2 the number of available configurational states in the system decreases until at the temperature T_2 the system possesses only one degree of freedom. This leads to

$$\frac{S_{conf}(T_2)}{n_x k T_2} = 0 = \phi\left(\frac{\epsilon_h}{kT_2}\right) + \lambda\left(\frac{\Delta\epsilon}{kT_2}\right) + \frac{1}{x}\ln\left[\{(z-2)x + 2\}\frac{(z-1)}{2}\right] \qquad (11.12)$$

where $\phi\left(\epsilon_h/kT\right) = \ln\left(\epsilon_h/S_0\right)^{z/2-1} + f_0/f_x \ln(f_0/S_0)$

and

$$\lambda\left(\frac{\Delta\epsilon}{kT}\right) = \frac{x-3}{x}\ln\left\{1 + (z-2)\exp(-\Delta\epsilon/kT) + (\Delta\epsilon/kT)\left[\frac{(z-2)\exp(-\Delta\epsilon/kT)}{1 + (z-2)\exp(-\Delta\epsilon/kT)}\right]\right\}$$

The fractions of unoccupied and occupied sites are f_0 and f_x respectively while S_0 is a function of f_0, f_x, and z. The main weaknesses of this theory are (a) that a chain of zero stiffness would have a T_g of 0 K and (b) that the T_g would be essentially independent of any intermolecular interactions. In spite of these limitations, various aspects of the behaviour of copolymers, plasticized polymers, and the chain length dependence of T_g, can be predicted in a reasonably satisfactory manner. The temperature T_2 is not of course an experimentally measurably quantity but is calculated to lie approximately 50 K below the experimental T_g and can be related to T_g on this basis.

ADAM–GIBBS THEORY
While the kinetic approach embodied in the WLF equation and the equilibrium treatment of the G-D theory have both been successful in their way, the one-sided aspect of each probably masks the fact that they are not entirely incompatible with one another. An attempt to reunite both channels of thought has been made by Adams and Gibbs who have outlined a molecular kinetic theory.

In this they relate the temperature dependence of the relaxation process to the temperature dependence of the size of a region, which is defined as a volume large enough to allow co-operative rearrangement to take place without affecting a neighbouring region. This "co-operatively rearranging region" is large enough to allow a transition to a new conformation, hence is determined by the chain conformation and by definition will equal the sample size at T_2 where only one conformation is available to each molecule. Evaluation of the temperature dependence of the size of such regions leads to an expression for the co-operative transition probability $W(T)$, which is simply the reciprocal of the relaxation time, and the final expression bears a gratifying resemblance to the WLF equation

$$-\log_{10} aT = -\log_{10} [W(T_0)/W(T)] = c_1(T-T_0)/ \{ c_2 + (T-T_0) \}, (11.13)$$

where $c_1 = 2.303\, S_c\Delta\mu/k\Delta c_p T_0 \ln(T_0/T_2)$,

and $c_2 = T_0 \ln (T_0/T_2)/ \{ 1 + \ln (T_0/T_2) \}$.

The temperature T_0 is the reference temperature of the WLF equation, T_2 is from the G-D theory, S_c is the macroscopic configurational entropy for the smallest region, Δc_p is the difference in specific heat capacity between the melt and the glass at T_g, and $\Delta\mu$ is the height of the potential energy barrier per monomer unit for a co-operative rearrangement.

Results plotted according to the WLF equation could be predicted also from the molecular kinetic equation and show that the two approaches are compatible. The Adam–Gibbs equations also lead to a value of $(T_g - T_2) = 55$ K, so the theory appears to resolve most of the differences between the kinetic and thermodynamic interpretations of the glass transition.

These theories point to the fundamental importance of T_2 as a true second-order transition temperature and to the experimental T_g as a temperature governed by the time scale of the measuring technique. The latter value has great practical significance, however, and is a parameter which is essential to the understanding of the physical behaviour of a polymer.

11.11 Dependence of T_g on molar mass

The value of T_g depends on the way in which it is measured but it is also found to be a function of the polymer chain length. At high molar masses the glass temperature is essentially constant when measured by any given method, but decreases as the molar mass of the sample is lowered. In terms of the simple free volume concept each chain end requires more free volume in which to move about than a segment in the chain interior. With increasing thermal energy the chain ends will be able to rotate more readily than the rest of the chain and the more chain ends a sample has the greater the contribution to the free volume when these begin moving, consequently the glass transition temperature is lowered. Bueche expressed this as

$$T_g(\infty) = T_g + K/M = T_g + (2\rho N_A \theta/\alpha M_n), \qquad (11.14)$$

where $T_g(\infty)$ is the glass temperature of a polymer with a very large molar mass, θ is the free volume contribution of one chain end and is 2θ for a linear polymer, ρ is the polymer density, N_A is Avogradro's constant, and α is the free volume expansivity defined as

$$\alpha = (\alpha_L - \alpha_G). \qquad (11.15)$$

Simha and Boyer found that, assuming the glass transition is an iso-free volume state, the expansivity α could be related to T_g by

$$(\alpha_L - \alpha_G)T_g = K_1, \tag{11.16}$$

where K_1 was a constant related somewhat later to the fractional free volume of the G-D theory by $K_1 = -f_0 \ln f_0$.

An expression similar to equation (11.14) was derived by Kanig using a thermodynamic approach in which the energy necessary for the formation of a "hole" was the dominating factor. This was also linear and, like equation (11.14), capable of describing data over a limited range of M_n. A more extensive correlation between T_g and M_n can be obtained using the G-D theory and if the assumption is made that T_2 is analogous to T_g, the variation with chain length is given by

$$\frac{x_n}{x_n - 2}\left[-\frac{f_0 \ln f_0}{(1-f_0)} + \frac{\ln 2x_n}{x_n} - \frac{(x_n-1)}{x_n}\right] = \frac{-\Delta\epsilon}{kT_g} \cdot \frac{2 \exp \beta}{(1 + 2 \exp \beta)} - \ln(1 + 2 \exp \beta). \tag{11.17}$$

where f_0 is the fractional free volume, k is the Boltzmann constant, $\beta = (-\Delta\epsilon/kT)$ and x_n is the chain length. Data for poly(methyl methacrylate) and poly(α-methyl styrene) shown in figure 11.9 have been compared with the theoretical expressions

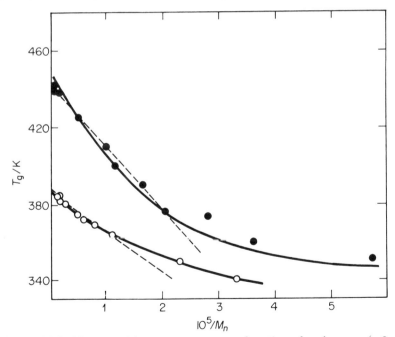

FIGURE 11.9. Glass transition temperature as a function of molar mass ($-\bullet-$) for poly(α-methyl styrene) [Cowie and Toporowski], and ($-\circ-$) for poly(methyl methacrylate) [Beevers and White]. Solid lines calculated from equation (11.17), broken lines from equation (11.14).

(11.14) and (11.17). The linear dependence embodied in equation (11.14) holds only at longer chain lengths whereas when values of $\Delta\epsilon = 4100$ and 5000 J mol^{-1} were substituted in equation (11.17), for the respective polymers, good agreement was obtained.

11.12 The glassy state

When a linear amorphous polymer is in the glassy state, the material is rigid and brittle because the flow units of the chain are co-operatively immobile and effectively frozen in position. The polymer sample is also optically transparent, as the chains are distributed in a random fashion and present no definite boundaries or discontinuities from which light can be reflected. An amorphous polymer in this state has been likened to a plate of frozen spaghetti. If a small stress is applied to a polymer glass, it exhibits a rapid elastic response resulting from purely local, bond angle, deformation. Consequently, although the modulus is high, specimen deformation is limited to about 1 per cent, due to the lack of glide planes in the disordered mass. This means that the sample has no way of dissipating a large applied stress, other than by bond rupture, and so a polymer glass is prone to brittle fracture.

Polymers do not, however, form perfectly elastic solids, as a limited amount of bond rotation can occur in the glass which allows slight plastic deformation; this makes them somewhat tougher than an inorganic glass.

General Reading

F. Bueche, *Physical Properties of Polymers.* Interscience Publishers Inc. (1962).
M. Gordon, *High Polymers.* Iliffe (1963).
P. Meares, *Polymers: Structure and Bulk Properties,* Chapter 10. Van Nostrand (1965).
A. V. Tobolsky and H. Mark, *Polymer Science and Materials,* Chapter 6. Wiley-Interscience (1971).

References

1. R. F. Boyer, *Rubber Che. Technol.,* **36,** 1303 (1963).
2. R. B. Beevers and E. F. T. White, *Trans. Farad. Soc.,* **56,** 744 (1960).
3. J. M. G. Cowie and P. M. Toporowski, *Europ. Polym. J.,* **4,** 621 (1968).
4. T. G. Fox and P. J. Flory, *J. Phys. Chem.,* **55,** 221 (1951).
5. A. J. Kovacs, *J. Polym. Sci.,* **30,** 131 (1958).
6. J. M. O'Reilly and F. E. Karasz, *J. Polym. Sci.,* **C,** No. **14,** 49 (1966).

CHAPTER 12

Mechanical Properties

12.1 Viscoelastic state

The fabrication of an article from a polymeric material in the bulk state, whether it be the moulding of a thermosetting plastic or the spinning of a fibre from the melt, involves deformation of the material by applied forces. Afterwards, the finished article is inevitably subjected to stresses, hence it is important to be aware of the mechanical and rheological properties of each material and understand the basic principles underlying their response to such forces.

In classical terms the mechanical properties of elastic solids can be described by Hooke's law, which states that an applied stress is proportional to the resultant strain, but is independent of the rate of strain. For liquids the corresponding statement is known as Newton's law, with the stress now independent of the strain, but proportional to the rate of strain. Both are limiting laws, valid only for small strains or rates of strain, and while it is essential that conditions involving large stresses, leading to eventual mechanical failure, be studied, it is also important to examine the response to small mechanical stresses. Both laws can prove useful under these circumstances.

In many cases, a material may exhibit the characteristics of both a liquid and a solid and neither of the limiting laws will adequately describe its behaviour. The system is then said to be in a *viscoelastic state*. A particularly good illustration of a viscoelastic material is provided by a silicone polymer known as "bouncing putty". If a sample is rolled into the shape of a sphere it can be bounced like a rubber ball, *i.e.* the rapid application and removal of a stress causes the material to behave like an elastic body. If on the other hand, a stress is applied slowly over a longer period the material flows like a viscous liquid so that the spherical shape is soon lost if left to stand for some time. Pitch behaves in a similar, if less spectacular, manner.

Before examining the viscoelastic behaviour of amorphous polymeric substances in more detail, some of the fundamental terms used will be defined.

12.2 Mechanical Properties

Homogeneous, isotropic, elastic materials possess the simplest mechanical properties and three elementary types of elastic deformation can be observed when such a body is subjected to (i) simple tension, (ii) simple shear, and (iii) uniform compression.

Simple tension. Consider a parallelepiped of length x_0 and cross-sectional area $A_0 = y_0 z_0$. If this is subjected to a balanced pair of tensile forces F, its length changes by an increment dx so that $x_0 + dx = x$. When dx is small, Hooke's law is obeyed, and the tensile *stress* σ is proportional to the tensile *strain* ϵ. The constant of proportionality is known as the *modulus,* and for elastic solids

$$\sigma = E\epsilon, \tag{12.1}$$

where E is Young's modulus.

The stress σ is a measure of the force per unit area (F/A), and the strain or elongation is defined as the extension per unit length, *i.e.* $\epsilon = (dx/x_0)$. It should be pointed out, however, that other definitions of strain will be met with in the literature, most notably, $\epsilon = \ln(x/x_0)$ is often called the *true strain,* while an expression arising from the kinetic theory of elasticity has the form

$$\epsilon = (1/3)\{(x/x_0) - (x_0/x)^2\}.$$

Of course, the extension dx will be accompanied by lateral contractions dy and dz, but although normally negative and equal, they can usually be assumed to be zero.

For an isotropic body, the change in length per unit length is related to the change in width per unit of length, such that

$$\nu_P = (dy/y_0)/(dx/x_0), \tag{12.2}$$

where ν_P is known as Poisson's ratio and varies from 0.5, when no volume change occurs, to about 0.2.

Simple shear. In simple shear the shape change is not accompanied by any change in volume. If the base of the body, shown shaded in the diagram, figure 12.1(b) is firmly fixed, a transverse force F applied to the opposite face is sufficient to cause a deformation dx through an angle θ. The shear modulus G is then given by the quotient of the shearing force per unit area and the shear per unit distance between shearing surfaces; and so

$$G = \sigma_s/\epsilon_s = (F/yz)/(dx/y) = F/A \tan \theta.$$

For very small shearing strains $\tan \theta \approx \theta$ and

$$G = F/A\theta. \tag{12.3}$$

(a)   (b)

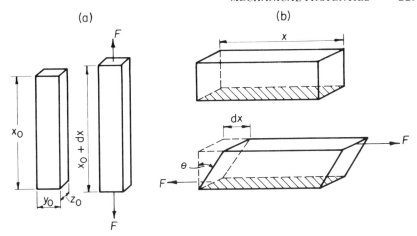

FIGURE 12.1. (a) Tensile stressing of a bar. (b) Shearing of a rectangular block, with balanced pair of forces F.

Both E and G depend on the shape of the specimen and it is usually necessary to define the shape carefully for any measurement.

Uniform compression. When a hydrostatic pressure $-p$ is applied to a body of volume V_0, causing a change in volume ΔV, a bulk modulus B can be defined as

$$B = -p/(\Delta V/V_0). \tag{12.4}$$

The quantity B is often expressed in terms of the compressibility which is the reciprocal of the bulk modulus. Similarly E^{-1} and G^{-1} are known as the tensile and shear compliances and given the symbols D and J respectively.

12.3 Interrelation of moduli

The relations given above pertain to isotropic bodies and for non-isotropic bodies the equations are considerably more complex. Polymeric materials are normally either amorphous, or partially crystalline with randomly oriented crystallites embedded in a disordered matrix. However, any symmetry possessed by an individual crystallite can be disregarded and the body as a whole is treated as being isotropic.

The various moduli can be related to each other in a simple manner, because an isotropic body is considered to possess only two independent elastic constants and so

$$E = 3B(1 - 2\nu_P) = 2(1 + \nu_P)G. \tag{12.5}$$

This indicates, that for an incompressible elastic solid, *i.e.* one having a Poisson ratio of 0.5, Young's modulus is three times larger than the shear modulus. These moduli have dimensions of pressure and typical values for several polymeric and non-polymeric materials can be compared at ambient temperatures in table 12.1.

TABLE 12.1. Comparison of various moduli for some common materials

Material	$E/\text{GN m}^{-2}$	ν_p	$G/\text{GN m}^{-2}$
Steel	220	0.28	85.9
Copper	120	0.35	44.4
Glass	60	0.23	24.4
Granite	30	0.30	15.5
Polystyrene	34	0.33	1.28
Nylon-6,6	20	–	–
Polyethylene	24	0.38	0.087
Natural Rubber	0.02	0.49	0.00067

The response of polymers to mechanical stresses can vary widely, and depends on the particular state the polymer is in at any given temperature.

12.4 Mechanical models describing viscoelasticity

A perfectly elastic material obeying Hooke's law behaves like a perfect spring. The stress-strain diagram is shown in figure 12.2(a), and can be represented in mechanical terms by the model of a weightless *spring* whose modulus of extension represents the modulus of the material.

The application of a shear stress to a viscous liquid on the other hand, is relieved by viscous flow, and for small values of σ_s can be described by Newton's law

$$\sigma_s = \eta \, d\epsilon_s / dt, \tag{12.6}$$

where η is the coefficient of viscosity and $(d\epsilon_s/dt)$ is the rate of shear sometimes denoted by $\dot{\gamma}$. As stress is now independent of the strain the form of the diagram changes and can be represented by a *dashpot* which is a loose fitting piston in a cylinder containing a liquid of viscosity η. (Figure 12.2(b).)

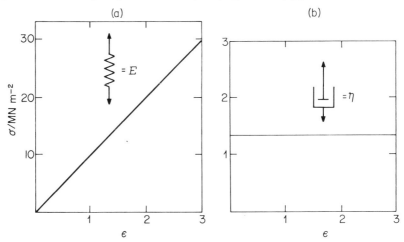

FIGURE 12.2. Stress-strain $(\sigma - \epsilon)$ behaviour of (a) spring of modulus E and (b) a dashpot of viscosity η.

Comparison of the two models shows that the spring represents a system storing energy which is recoverable, whereas the dashpot represents the dissipation of energy in the form of heat by a viscous material subjected to a deforming force. The dash pot is used to denote the retarded nature of the response of a material to any applied stress.

Because of their chain-like structure, polymers are not perfectly elastic bodies and deformation is accompanied by a complex series of long and short range co-operative molecular rearrangements. Consequently, the mechanical behaviour is dominated by viscoelastic phenomena, in contrast to materials such as metal and glass where atomic adjustments under stress are more localized and limited.

The Maxwell model. One of the first attempts to explain the mechanical behaviour of materials such as pitch and tar was made by James Clark Maxwell. He argued that when a material can undergo viscous flow and also respond elastically to a stress it should be described by a combination of both the Newton and Hooke laws. This assumes that both contributions to the strain are additive so that $\epsilon = \epsilon_{elast} + \epsilon_{visc}$. Expressing this as the differential equation leads to the equation of motion of a Maxwell unit

$$d\epsilon/dt = (1/G)(d\sigma/dt) + \sigma/\eta \qquad (12.7)$$

Under conditions of constant shear strain ($d\epsilon/dt = 0$) the relation becomes

$$d\sigma/dt + G\sigma/\eta = 0, \qquad (12.8)$$

and if the boundary condition is assumed that $\sigma = \sigma_0$ at zero time, the solution to this equation is

$$\sigma = \sigma_0 \exp(-tG/\eta), \qquad (12.9)$$

where σ_0 is the initial stress immediately after stretching the polymer. This shows that when a Maxwell element is held at a fixed shear strain, the shearing stress will relax exponentially with time. At a time $t = (\eta/G)$ the stress is reduced to $1/e$ times the original value and this characteristic time is known as the *relaxation time τ*.

The equations can be generalized for both shear and tension and G can be replaced by E. The mechanical analogue for the Maxwell unit can be represented by a combination of a spring and a dashpot arranged in series so that the stress is the same on both elements. This means that the total strain is the sum of the strains on each element as expressed by equation (12.7). A typical stress-strain curve predicted by the Maxwell model, is shown in figure 12.3(a). Under conditions of constant stress, a Maxwell body shows instantaneous elastic deformation first, followed by viscous flow.

Voigt–Kelvin Model. A second simple mechanical model can be constructed from the ideal elements by placing a spring and dashpot in parallel. This is known as a Voigt–Kelvin model. Any applied stress is now shared between the elements

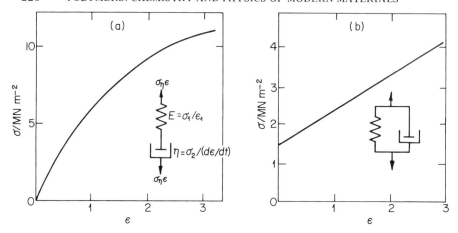

FIGURE 12.3. Stress-strain ($\sigma - \epsilon$) behaviour of two simple mechanical models, (a) the Maxwell model and (b) the Voigt-Kelvin model.

and each is subjected to the same deformation. The corresponding expression for strain is

$$\epsilon(t) = \sigma_0 J \{1 - \exp(t/\tau_R)\}. \qquad (12.10)$$

Here $\tau_R = (\eta/G)$ is known as the *retardation time* and is a measure of the time delay in the strain after imposition of the stress. For high values of the viscosity, the retardation time is long and this represents the length of time the model takes to attain $(1 - 1/e)$ or 0.632 of the equilibrium elongation.

Such models are much too simple to describe the complex viscoelastic behaviour of a polymer, nor do they provide any real insight into the molecular mechanism of the process, but in certain instances they can prove useful in assisting the understanding of the viscoelastic process.

12.5 Linear viscoelastic behaviour of amorphous polymers

A polymer can possess a wide range of material properties and of these the hardness, deformability, toughness, and ultimate strength, are amongst the most significant. Certain features, such as high rigidity (modulus) and impact strength, combined with low creep characteristics are desirable in a polymer if eventually it is to be subjected to loading. Unfortunately, these are conflicting properties, as a polymer with a high modulus and low creep response does not absorb energy by deforming easily, hence has poor impact strength. This means a compromise must be sought depending on the use to which the polymer will be put, and this requires a knowledge of the mechanical response in detail.

The early work on viscoelasticity was performed on silk, rubber, and glass, and it was concluded that these materials exhibited a "delayed elasticity" manifest in the observation, that the imposition of a stress resulted in an instantaneous

strain which continued to increase more slowly, with time. It is this delay
between cause and effect that is fundamental to the observed viscoelastic response
and the three major examples of this hysteresis effect are (1) *Creep,* where
there is a delayed strain response after the rapid application of a stress, (2) *Stress-
relaxation,* (section 12.7) in which the material is quickly subjected to a strain
and a subsequent decay of stress is observed, and (3) *Dynamic response* (section
12.9) of a body to the imposition of a steady sinusoidal stress. This produces
a strain oscillating with the same frequency as, but out of phase with, the stress.
For maximum usefulness, these measurements must be carried out over a wide
range of temperature.

CREEP

To be of any practical use, an object made from a polymeric material must be
able to retain its shape when subjected to even small tensions or compressions
over long periods of time. This dimensional stability is an important consideration in choosing a polymer to use in the manufacture of an item. No one wants a
plastic telephone receiver which sags after sitting in its cradle for several weeks,
or a car tyre that develops a flat spot if parked in one position for too long, or
clothes made from synthetic fibres which become baggy and deformed after
short periods of wear. Creep tests provide a measure of this tendency to deform
and are relatively easy to carry out.

Creep can be defined as a progressive increase in strain, observed over an
extended time period, in a polymer subjected to a constant stress. Measurements
are carried out on a sample clamped in a thermostat. A constant load is firmly
fixed to one end and the elongation is followed by measuring the relative
movement of two fiducial marks, made initially on the polymer, as a function
of time. To avoid excessive changes in the sample cross section, elongations are
limited to a few per cent and are followed over approximately three decades of
time.

The initial, almost instantaneous, elongation produced by the application of
the tensile stress is inversely proportional to the rigidity or modulus of the
material, *i.e.* an elastomer with a low modulus stretches considerably more than
a material in the glassy state with a high modulus. The initial deformation
corresponds to portion OA of the curve (figure 12.4), increment a. This rapid
response is followed by a region of creep, A to B, initially fast but eventually
slowing down to a constant rate represented by the section B to C. When the
stress is removed the instantaneous elastic response OA is completely recovered
and the curve drops from C to D, *i.e.* the distance $a' = a$. There follows a
slower recovery in the region D to E which is never complete, falling short of
the initial state by an increment $c' = c$. This is a measure of the viscous flow
experienced by the sample and is a completely non-recoverable response. If the
tensile load is enlarged, both the elongation and the creep rate increase, so
results are usually reported in terms of the *creep compliance D(t)*, defined as

the ratio of the relative elongation y at time t to the stress so that

$$D(t) = yE/\sigma \qquad (12.11)$$

At low loads $D(t)$ is independent of the load.

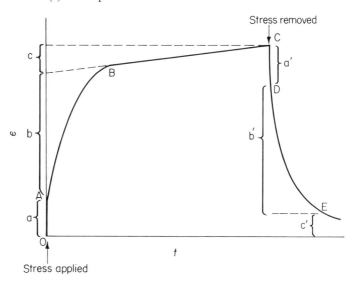

FIGURE 12.4. Schematic representation of a creep curve: a, initial elastic response; b, region of creep; c, irrecoverable viscous flow. This curve can be represented by the four element model shown in figure 12.5.

This idealized picture of creep behaviour in a polymer has its mechanical equivalent constructed from the springs and dashpots described earlier. The changes a and a' correspond to the elastic response of the polymer and so we can begin with a Hookean spring. The Voigt–Kelvin model is embodied in equation (12.11) and this reproduces the changes b and b'. The final changes c and c' represent viscous flow and can be represented by a dashpot so that the whole model is a four element model – figure 12.5.

The behaviour can be explained in the following series of steps. In diagram (i) the system is at rest. The stress σ is applied to spring E_1 and dashpot η_3; it is also shared by E_2 and η_2 but in a manner which varies with time. In diagram (ii), representing zero time, the spring E_1 extends by an amount $\sigma/E_1 (= a)$. This is followed by a decreasing rate of creep with a progressively increasing amount of stress being carried by E_2 until eventually none is carried by η_2 and E_2 is fully extended – diagram (iii). Such behaviour is described by

$$\epsilon(t) = (\sigma_0/E_2) \left\{ 1 - \exp(- t/\tau_R) \right\} \qquad (12.12)$$

where the retardation time τ_R provides a measure of the time required for E_2 and η_2 to reach 0.632 of their total deformation. A considerably longer time is required for complete deformation to occur. When spring E_2 is fully extended the creep attains a constant rate corresponding to movement in the dashpot η_3. Viscous flow continues and the dashpot η_3 is deformed until the stress is removed. At that time, E_1 retracts quickly along section a' and a period of recovery ensues (b'). During this time spring E_2 forces the dashpot plunger in η_2 back to its original position. As no force acts on η_3 it remains in the extended state, and corresponds to the non-recoverable viscous flow; region $c' = \sigma t/\eta_3$. The system is then as shown in diagram (v). In practice, a substance possesses a large number of retardation times which can be expressed as a distribution function $L_1(\tau)$ where

$$L_1(\tau) = \mathrm{d}\{D(t) - (t/\eta)\}/\mathrm{d} \ln t. \qquad (12.13)$$

To the first approximation, this is estimated from a plot of creep compliance against $\log_e t$, and (t/η) is the contribution from viscous flow.

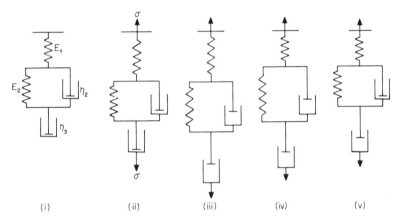

FIGURE 12.5. Use of mechanical models to describe the creep behaviour of a polymeric material.

STRESS-STRAIN MEASUREMENTS

The data derived from stress-strain measurements on thermoplastics are important from a practical viewpoint, providing as they do, information on the modulus, the brittleness, and the ultimate and yield strengths of the polymer. By subjecting the specimen to a tensile force applied at a uniform rate and measuring the resulting deformation, a curve of the type shown in figure 12.6 can be constructed.

The shape of such a curve is dependent on the rate of testing, consequently, this must be specified if a meaningful comparison of data is to be made. The initial portion of the curve is linear and the tensile modulus E is obtained from its slope. The point L represents the stress beyond which a brittle material will fracture, and the area under the curve to this point is proportional to the energy required for brittle fracture. If the material is tough no fracture occurs, and

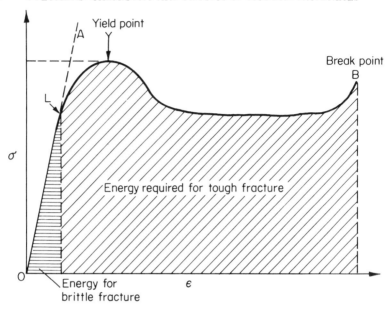

FIGURE 12.6. Idealized stress-strain curve. The slope of line OA is a measure of the true modulus.

the curve then passes through a maximum or inflection point Y, known as the yield point. Beyond this, the ultimate elongation is eventually reached and the polymer breaks at B. The area under this part of the curve is the energy required for tough fracture to take place.

EFFECT OF TEMPERATURE ON STRESS-STRAIN RESPONSE
Polymers such as polystyrene and poly(methyl methacrylate) with a high E at ambient temperatures fall into the category of hard brittle materials which break before point Y is reached. Hard tough polymers can be typified by cellulose acetate and several curves measured at different temperatures are shown in figure 12.7(a). Stress-strain curves for poly(methyl methacrylate) are also shown for comparison (figure 12.7(b)).

It can be seen that the effect of temperature on the characteristic shape of the curve is significant. As the temperature increases both the rigidity and the yield strength decrease while the elongation generally increases. For celullose acetate there is a transformation from a hard brittle state below 273 K to a softer but tougher type of polymer at temperatures above 273 K. For poly-(methyl methacrylate) the hard brittle characteristics are retained to much higher temperature, but it eventually reaches a soft tough state at about 320 K. Thus if the requirements of high rigidity and toughness are to be met the temperature is important. Cellulose acetate meets these requirements if used at 298 K more satisfactorily than when used at 350 K where the modulus is

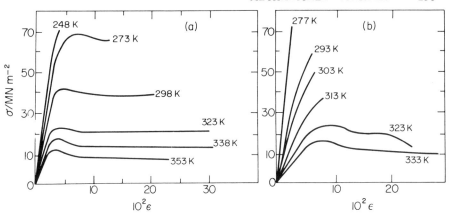

FIGURE 12.7. Influence of temperature on the stress-strain response of (a) cellulose acetate and (b) poly(methyl methacrylate). (From data by Carswell and Nason.)

smaller and the ability to absorb energy, represented by the area under the curve, is also lower.

12.6 Boltzmann superposition principle

If a Hookean spring is subjected to a series of incremental stresses at various times, the resulting extensions will be independent of the loading or past history of the spring. A Newtonian dashpot also behaves in a predictable manner. For viscoelastic materials the response to mechanical testing is time dependent, but the behaviour at any time can be predicted by applying a superposition principle proposed by Boltzmann. This can be illustrated by a creep test using a simple Voigt–Kelvin model with a single retardation time τ_R, placed initially under a stress σ_0 at time t_0. If after times t_1, t_2, t_3, ... the system is subjected to additional stresses σ_1, σ_2, σ_3, ... then the principle states that the creep response of the system can be predicted simply by summing the individual responses from each stress increment. Thus if the stress alters continually, the summation can be replaced by an integral, and (σ_n by a continually varying function), so that at time t^* when the stress $\sigma(t^*)$ existed, the strain is given by

$$\epsilon(t^*) = \int_0^{t^*} \frac{\mathrm{d}\sigma(t^*)}{\mathrm{d}t^*} \, \phi(t^* - t_n) \, \mathrm{d}t. \tag{12.14}$$

The principle has been applied successfully to the tensile creep of amorphous and rubber-like polymers, but it is not too successful if appreciable crystallinity exists in the sample. Graphical representation of the principle is shown in figure 12.8.

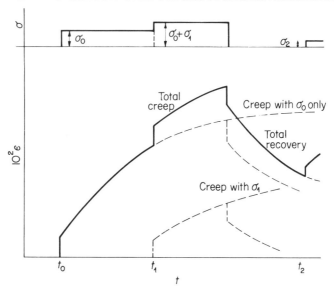

FIGURE 12.8. Application of the Boltzmann superposition principle to a creep experiment.

12.7 Stress-relaxation

Stress-relaxation experiments involve the measurement of the force required to maintain the deformation produced initially by an applied stress as a function of time. Stress-relaxation tests are not performed as often as creep tests because many investigators believe they are less readily understood. The latter point is debatable and it may only be that the practical aspects of creep measurements are simpler. As will be shown later, all the mechanical parameters are in theory interchangeable, and so all such measurements will contribute to the understanding of viscoelastic theory. While stress-relaxation measurements are useful in a general study of polymeric behaviour, they are particularly useful in the evaluation of antioxidants in polymers, especially elastomers, because measurements on such systems are relatively easy to perform and are sensitive to bond rupture in the network.

Experimental stress-relaxation technique. In a stress-relaxation experiment, the sample under study is deformed by a rapidly applied stress. As the stress is normally observed to reach a maximum as soon as the material deforms and then decreases thereafter, it is necessary to alter this continually in order to maintain a constant deformation or measure the stress that would be required to accomplish this operation.

The apparatus used varies in complexity with the physical nature of the sample, being simplest for an elastomer and becoming more sophisticated when the polymer is more rigid. One type of experimental set up is shown in figure 12.9. The sample is fixed in position by means of clamps, one being attached to a spring beam above and the other to an adjustable rod R below. A stress is applied to the sample by rapidly pulling rod R downwards and clamping it in position. This causes the beam to bend and the displacement is measured by means of a strain gauge or a differential transformer. The beam deflection is then fed to a recorder and a trace of stress against time is obtained.

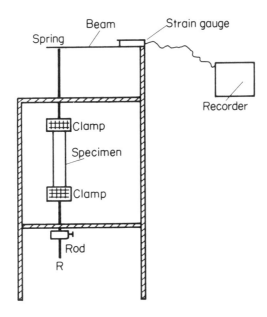

FIGURE 12.9. Simple apparatus to measure the stress-relaxation of a polymer.

The results are expressed as a relaxation modulus $E_r(t)$ which is a function of the time of observation. Typical data for polyisobutylene are shown in figure 12.10, where the logarithm of the relaxation modulus $\log E_r(t)$ is plotted against $\log t$. From the curves it can be seen that there is a rapid change in $\log E_r(t)$ over a narrow range of temperature corresponding to the glass transition.

Again a simple model with a single relaxation time is too crude, and the stress-relaxation modulus $E_r(t)$ is better represented by

$$E_r(t) = \int_0^\infty H(\tau) \exp(-t/\tau)\, \mathrm{d}(\ln \tau), \qquad (12.15)$$

where $H(\tau)$ is the distribution function of relaxation times. This is suitable for a linear polymer but requires the additional term E_∞ if the material is crosslinked.

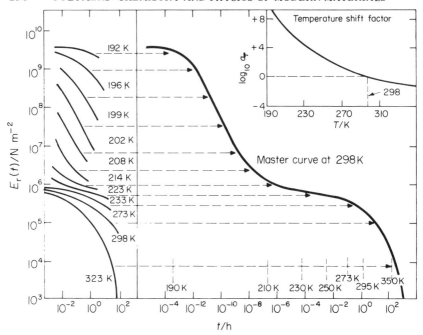

FIGURE 12.10. Illustration of the time-temperature superposition principle using stress-relaxation data for polyisobutylene. Curves are shifted along the axis by an amount represented by a_T as shown in the insert. The reference temperature in this instance is 298 K. (Adapted from Castiff and Tobolsky.)

12.8 Time-temperature superposition principle

A curve of the logarithm of the modulus against time and temperature provides a particularly useful description of the behaviour of a polymer and allows one to estimate, among other things, either the relaxation or retardation spectrum.

The practical time scale for most stress-relaxation measurements ranges from 10^1 to 10^6 s but a wider range of temperature is desirable. Such a range can be covered relatively easily by making use of the observation, first made by Leaderman, that for viscoelastic materials time is equivalent to temperature. A composite isothermal curve covering the required extensive time scale can then be constructed from data collected at different temperatures. This is accomplished by translation of the small curves along the log t axis until they are all super-imposed to form a large composite curve. The technique can be illustrated using data for polyisobutylene at several temperatures. An arbitrary temperature T_0 is first chosen to serve as a reference which in the present case is 298 K. As values of the relaxation modulus $E_r(t)$ have been measured at widely differing temperatures, they must be corrected for changes in the sample density with temperature to give a reduced modulus, where ρ and ρ_0 are the polymer densities at T and T_0 respectively. This correction is small and can often be neglected.

$$[E_r(t)]_{red} = (T_0\rho_0/T\rho)E_r(t). \tag{12.16}$$

Each curve of reduced modulus is shifted with respect to the curve at T_0 until all fit together forming one master curve. The curve obtained at each temperature is shifted by an amount

$$(\log t - \log t_0) = \log (t/t_0) = \log a_T \tag{12.17}$$

The parameter a_T is the shift factor and is positive if the movement of the curve is to the left of the reference and negative for a move to the right. The shift factor is a function of temperature only and decreases with increasing temperature, it is, of course, unity at T_0. The superposition principle can also be applied to creep data. Curves exhibiting the creep behaviour of polymers at different temperatures can be compared by plotting $D(t)T$ against $\log t$. This reduces all the curves at various temperatures to the same shape but displaced along the $\log t$ axis. Superposition to form a master curve is readily achieved by movement along the $\log t$ axis, where the shift factor a_T has the same characteristics as for the relaxation data. This shift factor has also been defined as the ratio of relaxation or retardation times at the temperatures T and T_0, i.e.

$$a_T = \tau/\tau_0 = (\eta/\eta_0)(T_0\rho_0/T\rho), \tag{12.18}$$

and is related to the viscosities. If the viscosities obey the Arrhenius equation, then by neglecting the correction factor, we can express a_T in an exponential form as

$$a_T = \exp b(1/T - 1/T_0) \tag{12.19}$$

or $\qquad \log_{10} a_T = -b(T - T_0)/2.303\, TT_0, \tag{12.20}$

where b is a constant.

This equation is very similar in form to the WLF equation,

$$\log_{10} a_T = -a_1(T - T_0)/(a_2 + T - T_0) \tag{12.21}$$

For polyisobutylene, the shift factor a_T can be predicted if $T_0 = (T_g + 45\text{ K})$ is used with $a_1 = 8.86$ and $a_2 = 101.6$ K. As outlined in Chapter 11, the reference temperature is often chosen to be T_g with $a_1 = 17.44$ and $a_2 = 51.6$ K, from which a_T can be calculated for various amorphous polymers.

The superposition principle can be used to predict the creep and relaxation behaviour at any temperature if some results are already available, with the proviso that the most reliable predictions can be made for interpolated temperatures rather than long extrapolations.

12.9 Dynamic mechanical response
The foregoing techniques for measuring mechanical properties are transient or non-periodic methods and typically cover time intervals of up to 10^6 s. For information relating to short times, the response of the polymer to stresses which vary with time can be gauged using dynamic mechanical testing methods.

These reflect the delayed reaction of a material to a stress and require that the sample be subjected to a sinusoidally varying stress of angular frequency ω. If the polymer is treated as a classical damped harmonic oscillator, both the elastic modulus and the damping characteristics can be obtained. Elastic materials convert mechanical work into potential energy which is recoverable; for example an ideal spring, if deformed by a stress, stores the energy and uses it to recover its original shape after removal of the stress. No energy is converted into heat during the cycle and so no damping is experienced.

Liquids on the other hand flow if subjected to a stress; they do not store the energy but dissipate it almost entirely as heat and thus possess high damping characteristics. Viscoelastic polymers exhibit both elastic and damping behaviour. Hence if a sinusoidal stress is applied to a linear viscoelastic material, the resulting stress will also be sinusoidal, but will be out of phase when there is energy dissipation or damping in the polymer.

Harmonic motion of a Maxwell element. The application of a sinusoidal stress to a Maxwell element produces a strain with the same frequency as, but out of phase with, the stress. This can be represented schematically in figure 12.11

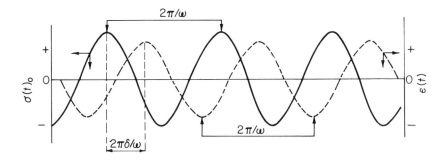

FIGURE 12.11. Harmonic oscillation of a Maxwell model, with the solid line representing the stress and the broken line the strain curve. This shows the lag in the response to an applied stress.

where δ is the phase angle between the stress and the strain. The resulting strain can be described in the terms of its angular frequency ω and the maximum amplitude ϵ_0 using complex notation, by

$$\epsilon^* = \epsilon_0 \exp(i\omega t), \qquad (12.22)$$

where $\omega = 2\pi\nu$, the frequency is ν and $i = -1^{1/2}$. The relation between the alternating stress and strain is written as

$$\sigma^* = \epsilon^* E^*(\omega), \qquad (12.23)$$

where $E^*(\omega)$ is the frequency dependent complex dynamic modulus defined by

$$E^*(\omega) = E'(\omega) + iE''(\omega). \qquad (12.24)$$

This shows that $E^*(\omega)$ is composed of two frequency dependent components; $E'(\omega)$ is the real part in phase with the strain called the *storage modulus,* and $E''(\omega)$ is the *loss modulus* defined as the ratio of the component $90°$ out of phase with the stress to the stress itself. Hence $E'(\omega)$ measures the amount of stored energy and $E''(\omega)$, sometimes called the imaginary part, is actually a real quantity measuring the amount of energy dissipated by the material.

The response is often expressed as a complex dynamic compliance

$$D^*(\omega) = D'(\omega) - iD''(\omega), \qquad (12.25)$$

especially if a generalized Voigt model is used. For a Maxwell model

$$\sigma^*/\epsilon^* = E\omega^2\tau^2/(1 + \omega^2\tau^2) + iE\omega\tau/(1 + \omega^2\tau^2). \qquad (12.26)$$

In more realistic terms, there is a distribution of relaxation times and a continuous distribution function can be derived, if required.

The damping in the system or the energy loss per cycle can be measured from the "loss tangent" $\tan \delta$. This is a measure of the internal friction and is related to the complex moduli by

$$\tan \delta = 1/\omega\tau = E''(\omega)/E'(\omega) = D''(\omega)/D'(\omega). \qquad (12.27)$$

The onset of molecular motion in a polymer sample is reflected in the behaviour of E' and E''. A schematic diagram (figure 12.12) of the variation of E' and E'' as a function of ω, assuming only a single value for τ in the model, shows that a maximum in the loss angle is observed where $\omega = 1/\tau$. This represents a transition point such as T_g, T_m, or some other region where significant molecular motion occurs in the sample. The maximum is characteristic of the dynamic method as the creep and relaxation techniques merely show a change in the modulus level.

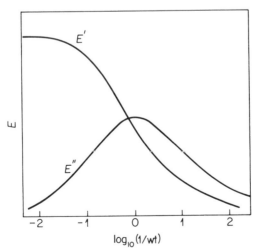

FIGURE 12.12. Behaviour of E' and E'' as a function of the angular frequency for a system with a single relaxation time.

12.10 Experimental methods

There are three main experimental approaches for measuring the dynamic mechanical properties of a sample, (a) free vibration, (b) forced vibration – resonance, (c) forced vibration – non-resonance. The mechanical response is usually determined at low frequencies and over as wide a temperature range as possible and examples of each are described in the following section.

TORSIONAL PENDULUM – FREE VIBRATION

A study of the mechanical damping and shear modulus under free vibration can be made using a torsional pendulum. The specimen is firmly fixed at one end and the other end is clamped to a disc, with a large moment of inertia, which can move freely. As the polymer sample should not be under a tensile stress, the suspension wire supporting the disc is passed over a pulley and the weight of the disc and sample are counterbalanced by loading the end. If the disc is subjected to an angular displacement and then released, the sample will twist backwards and forwards about the vertical axis. The oscillations stimulated in the sample are picked up by an arm attached to the rigidly fixed end held in torsion bars, and transmitted to a recorder by a linear variable differential transformer. The sample movements are traced as a series of oscillations whose frequency is a function of the physical state of the sample. The period of oscillation P is taken as the distance between adjacent maxima or minima and the amplitude A is a measure of the height from one minimum to the preceding maximum. The exponential decay of the amplitude along the axis provides an

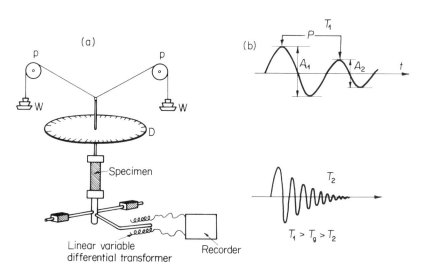

FIGURE 12.13. (a) Schematic diagram of a torsional pendulum in which the weight of the disc D is counterbalanced by weights W suspended over pulleys p. (b) Typical curves from a sample at T_2 below and at T_1 above its glass transition temperature.

indication of the mechanical damping. At a temperature $T_1 > T_g$ the sample absorbs most of the energy and damping is high whereas at a much lower temperature $T_2 < T_g$ the material tends to store the energy and mechanical damping is considerably lower.

A quantitative measure of the damping is provided by the logarithmic decrement Δ defined as the logarithmic decrease in amplitude per cycle. It is calculated from the ratio of amplitudes of any two successive oscillations using the relation

$$\Delta = \ln(A_1/A_2) = \ln(A_2/A_3) = \ldots\ldots = \ln(A_n/A_{n+1}). \qquad (12.28)$$

The shear modulus can also be derived from the data, being inversely proportional to the square of the period $G = KI/P^2$, where K is a factor depending on the shape and the size of the sample and I is the polar moment of inertia.

The method can cover the complete range of moduli encountered in polymeric systems but is confined to a relatively narrow frequency range of 0.01 to 10 Hz.

VIBRATING REED – RESONANCE

For resonance forced vibration measurements a sample in the form of a thin strip is clamped firmly at one end leaving the other end free. The clamped end of the system is then vibrated laterally at a given frequency ν and the amplitude of the vibration induced at the free end of the sample is recorded. A range of frequencies wide enough to ensure that it encompasses the resonant frequency

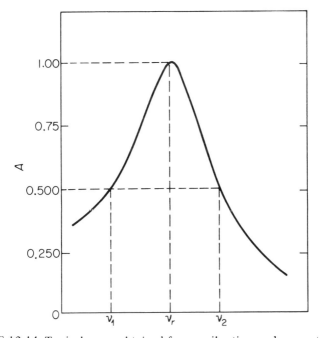

FIGURE 12.14. Typical curve obtained from a vibrating reed apparatus.

of the sample ν_r is then examined. The resonant frequency is detected as the maximum of a graph of amplitude against frequency. The results provide information on the elastic modulus E since it is related to the square of the resonance frequency by

$$E = cL^4 \rho \nu_r^2 / D^2 \qquad (12.29)$$

where c is a numerical constant, L is the free length of the sample, D is its thickness, and ρ is the sample density.

If the amplitudes are expressed as ratios of the amplitude to the maximum amplitude, then damping is measure d from the half-width h of the curve, *i.e.*

$$h = (\nu_2 - \nu_1)/\nu_r. \qquad (12.30)$$

This technique is not as useful as the torsional pendulum but covers the higher frequency range 10 to 10^3 Hz.

FORCED VIBRATION – NON-RESONANCE
Several types of instrument can be used for this type of test, and these are usually limited to measurements on rigid polymers or rubbers. One such instrument is shown in the block diagram 12.15. The sample C is attached firmly at each end to a strain gauge; one of these is a force transducer measuring the applied sinusoidal force and the other records the sample deformation. A sinusoidal tensile stress of a given frequency can be generated in the vibrator A and if the electrical vectors from the force and displacement are represented by α_1 and α_2 then by satisfying the condition $|\bar{\alpha}_1| = |\bar{\alpha}_2| = 1$ the tangent of the phase angle δ between the stress and the strain may be calculated from

$$\bar{\alpha}_1 - \bar{\alpha}_2 = 2\sin(\delta/2) \approx \tan\delta. \qquad (12.31)$$

This operation of adjustment followed by subtraction of the electrical vectors is performed directly in the recording circuit.

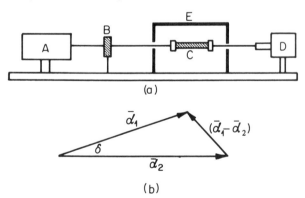

FIGURE 12.15. (a) Block diagram of apparatus for measuring the dynamic mechanical response using a non-resonance technique, (b) Vector diagram showing relation between α and δ.

The complex elastic modulus E^* is given by

$$E^* = FL/\Delta LA, \tag{12.32}$$

where F is the amplitude of the tensile force, A is the sample cross-sectional area, L is the sample length, and ΔL is the amplitude of elongation. Tensile storage and loss moduli E' and E'' follow from $E' = E^* \cos \delta$ and $E'' = E^* \sin \delta$.

12.11 Correlation of mechanical damping terms

The several practical methods described express the damping and moduli in slightly different forms but these can all be interrelated quite simply. In general, one can select a dissipation factor or loss tangent derived from the ratio (G''/G') or (E''/E') to represent the energy conversion per cycle. This leads to the equivalent forms

$$G''/G' = \Delta/\pi = (1/\pi n) \ln(A_1/A_n), \tag{12.33}$$

$$E''/E' = (1/\sqrt{3})(\nu_2 - \nu_1)/\nu_r, \tag{12.34}$$

and

$$E''/E' = \tan \delta. \tag{12.35}$$

To a first approximation it is also possible to write

$$(E''/E') = (G''/G')$$

thereby allowing use of the data from either type of measurement to characterize the sample. It should also be noted that if complex moduli are used the corresponding complex compliances are given by $(G''/G') = (J''/J')$. Moduli can also be related to the viscosity, $G' = \omega\eta''$ and $G'' = \omega\eta'$ where η is known as the dynamic viscosity.

The approximations $G \approx G'$ and $E \approx E'$ can be made when damping is low, and the absolute value for the modulus $|G|$ or $|E|$ can be related to the complex components by $|E| = \{(E')^2 + (E'')^2\}^{1/2}$. A similar expression holds for $|G|$.

12.12 A molecular theory for viscoelasticity

So far the interpretation of viscoelastic behaviour has been largely phenomenological, relying on the application of mechanical models to aid the elucidation of the observed phenomena. These are, at best, no more than useful physical aids to illustrate the mechanical response and suffer from the disadvantage that a given process may be described in this way using more than one arrangement of springs and dashpots. In an attempt to gain a deeper understanding on a molecular level, Rouse, Zimm, Bueche, and others have attempted to formulate a theory of polymer viscoelasticity based on a chain model consisting of a series of sub-units. Each sub-unit is assumed to behave like an entropy spring and is expected to be large enough to realize a Gaussian distribution of segments (*i.e.* > 50 carbon atoms). This approach, although still somewhat restrictive has led to reasonable predictions of relaxation and retardation spectra.

One starts with a single isolated chain and the assumption that it exhibits both viscous and elastic behaviour. If the chain is left undisturbed it will also adopt the most notable conformation or segmental distribution, so that, with the exception of high frequencies, the observed elasticity is predominantly entropic. Thus the application of a stress to the molecule will cause distortion, by altering the equilibrium conformation to a less probable one, resulting in a decrease in the entropy and a corresponding increase in the free energy of the system. When the stress is removed the chain segments will diffuse back to their unstressed positions even though the whole molecule may have changed its spatial position in the meantime. If on the other hand, the stress is maintained, strain relief is sought by converting the excess free energy into heat, thereby stimulating the thermal motion of the segments back to their original positions. *Stress-relaxation* is then said to have occurred. For a chain molecule composed of a large number of segments, movement of the complete molecule depends on the co-operative movement of all the segments, and as stress-relaxation depends on the number of ways the molecule can regain its most probable conformation, each possible co-ordinated movement is treated as a mode of motion with a characteristic relaxation time. For simplicity we can represent the polymer as in figure 12.16.

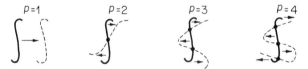

FIGURE 12.16. First four normal modes of movement of a flexible polymer molecule.

The first mode $p = 1$ represents translation of the molecule as a whole and has the longest relaxation time τ_1 because the maximum number of co-ordinated segmental movements are involved. The second mode $p = 2$ corresponds to the movement of the chain ends in opposite directions; for $p = 3$, both chain ends move in the same direction, but the centre moves in the opposite direction. Higher modes 4, 5 . . . m follow involving a progressively decreasing degree of co-operation for each succeeding mode and correspondingly lower relaxation times τ_p. This means that a single polymer chain possesses a wide distribution of relaxation times. Using this concept, Rouse considered a molecule in dilute solution under sinusoidal shear and derived the relations

$$\eta' = (G''/\omega) = \eta_s + nkT \sum_{p=1}^{m} \tau_p/(1 + \omega^2 \tau_p^2), \qquad (12.36)$$

$$\eta'' = (G'/\omega) = (nkT/\omega) \sum_{p=1}^{m} \omega^2 \tau_p^2/(1 + \omega^2 \tau_p^2), \qquad (12.37)$$

$$\tau_p = 6(\eta - \eta_s)/(\pi^2 p^2 nkT), \qquad (12.38)$$

where η and η_s are the viscosities of the solution and the solvent respectively, n is the number of molecules per unit volume, k is the Boltzmann constant, and ω is the angular frequency of the applied stress which is zero for steady flow.

These equations are strictly applicable only to dilute solutions of non-draining monodisperse coils, but can be extended to undiluted polymers above their glass temperature if suitably modified. This becomes necessary when chain entanglements begin to have a significant effect on the relaxation times. The undiluted system is represented as a collection of polymer segments dissolved in a liquid matrix composed of other polymer segments and η_s can be replaced by a monomeric frictional coefficient ζ_0. This provides a measure of the viscous resistance experienced by a chain and is characteristic of a given polymer at a particular temperature. The continuous relaxation and retardation spectra calculated from the Rouse theory are

$$H(\tau) = (\rho N_A / 2\pi M)(r_0^2 N k T \zeta_0 / 6\tau)^{1/2}, \tag{12.39}$$

and
$$L(\tau) = (2M/\pi \rho N_A)(6\tau_R / r_0^2 N k T \zeta_0)^{1/2}, \tag{12.40}$$

where r_0^2 is the unperturbed mean square end-to-end distance of a chain of molar mass M and density ρ containing N monomer units. The equations predict linearity in the plots $\log H(\tau)$ and $\log L(\tau_R)$ against $\log \tau$ with slopes of $-\frac{1}{2}$ and $+\frac{1}{2}$ respectively. Comparison with experimental results for poly(methyl acrylate) shows validity only for longer values of the relaxation and retardation times.

The Rouse model only pertains to the region covering intermediate τ values. The reason for this lies in the response of a polymer to an alternating stress. At low frequencies Brownian motion can relieve the deformation caused by the stress before the next cycle takes place, but as the frequency increases the conformational change begins to lag behind the stress and energy is not only dissipated but stored as well. Finally at very high frequencies only enough time exists for bond deformation to occur. As it was stipulated that each segment be long enough to obey Gaussian statistics, short relaxation times may not allow a segment sufficient time to rearrange and regain this distribution. Thus the contribution from short segments to the distribution functions tends to be lost and deviations from the theoretical represent departure from ideal Gaussian behaviour.

This approach to viscoelastic theory is reasonably successful in the low modulus regions but it requires considerable modification if the high modulus and rubbery plateau regions are to be described.

General Reading

F. Bueche, *Physical Properties of Polymers.* Interscience Publishers Inc. (1962).
J. D. Ferry, *Viscoelastic Properties of Polymers.* John Wiley and Sons (1970).
P. Meares, *Polymers: Structure and Bulk Properties,* Chapters 9 and 11. Van Nostrand (1965).
L. E. Nielsen, *Mechanical Properties of Polymers.* Reinhold Publishing Corp. (1962).
A. V. Tobolsky, *Properties and Structure of Polymers.* Interscience Publishers Inc. (1960).

A. V. Tobolsky and H. Mark, *Polymer Science and Materials*, Chapter 10. Wiley-Interscience (1971).

I. M. Ward, *Mechanical Properties of Solid Polymers*. John Wiley and Sons (1971).

References

1. E. Castiff and A. V. Tobolsky, *J. Colloid Sci.*, **10**, 375 (1955).
2. T. S. Carswell and H. K. Nason, ASTM Symposium on Plastics, Philadelphia (1944).

CHAPTER 13

The Elastomeric State

13.1 General introduction

Most materials, when stressed, exhibit a limited elastic region where the material regains its original dimensions if the stress is removed. As the resulting strain is related to the extent of movement of atoms from their equilibrium conditions, substances such as metals and glass have elastic limits rarely exceeding 1 per cent because atomic adjustments are localized. For long chain polymers, under certain conditions, the situation is different; the extensive covalent bonding between the atoms to form chains allows considerable deformation, which is accompanied by long and short range co-operative molecular rearrangement arising from the rotation about chain bonds.

One of the first materials found to exhibit a sizeable elastic region was a natural substance obtained from the tree *Hevea brasiliensis,* now known to us as *cis*-polyisoprene, or more commonly referred to as rubber. A large number of polymers, with rubber-like characteristics at ambient temperatures, are now available and it is preferable to call the general group of polymers, elastomers. Elastomers possess several significant characteristics:

(1) The materials are above their glass temperature;
(2) They possess the ability to stretch and retract rapidly;
(3) They have high modulus and strength when stretched;
(4) The polymers have a low or negligible crystalline content;
(5) The molar mass is large enough for network formation or they must be readily crosslinked.

The most important factor is of course the T_g as this determines the range of temperatures where elastomeric behaviour is important and defines its lower limiting temperature. Hence polymers with T_g below ambient may be useful elastomers, if they are essentially amorphous, whereas this is unlikely if T_g is in excess of 400 K. Environmental temperature is, of course, an important factor, and on a different planet, much colder than our own, latex rubber could prove to be a good glassy material, while on a hot planet, perspex could well be adapted for use an as elastomer.

One of the features of an elastomer is its ability to deform elastically by elongating up to several hundred per cent, but as we have seen in chapter 11 chain slippage occurs under prolonged tension and the sample deforms. This flow under stress can be greatly reduced by introducing crosslinks between the chains. In effect, these crosslinks act as anchors or permanent entanglements and prevent the chains slipping past each other. The process of crosslinking is known generally as vulcanization and the resultant polymer is a network of interlinked molecules now capable of maintaining an equilibrium tension. This is most important as it changes the properties of an elastomer to a marked degree and extends the usefulness of the polymer as a material.

NATURAL RUBBER (NR)
Naturally occurring rubber is a linear polymer of isoprene units linked 1,4

$$+CH_2-CH=C-CH_2 +_n$$
$$\qquad\qquad\qquad\qquad |$$
$$\qquad\qquad\qquad\qquad CH_3$$

and because the chain is unsaturated, two forms are found. Natural rubber is the *cis* form; it has low crystallinity, T_g = 200 K and T_m = 301 K,

whereas the *trans* form, called gutta percha or balata, is of medium crystallinity with T_g = 200 K and T_m = 347 K.

The remarkable effect of *cis* and *trans* isomerism on the properties is well illustrated in the case of the polyisoprenes. The more extended all *trans* form of balata allows the polymer to develop a greater degree of crystallinity and order. This is reflected in its hard tough consistency which makes it suitable, in the vulcanized state, for golfball covers. X-ray diffraction reveals two forms of gutta percha: an α-form with a molecular repeat distance of 0.88 nm, slightly larger than a *cis* unit and a more compact β-form with a repeat distance of 0.47 nm.

The chains of the *cis*-polyisoprene are more easily rotated than their *trans* counterparts and as a result the molecules prefer to coil up into a compact conformation. The viscoelastic behaviour and long range elasticity arise from this random arrangement of long, freely moving chains, and recoverable deformations of up to 1000 per cent can be observed. In the raw state natural rubber is a tacky substance rather difficult to handle, has poor abrasion resistance, and is sensitive to oxidative degradation. It can be used for making crepe soles and adhesives but it is immensely improved if vulcanized, a process which cross-links the rubber. Vulcanization enhances resistance to degradation and increases both the tensile strength and elasticity.

13.2 Experimental vulcanization

The process of vulcanization was discovered independently by Goodyear (1839) in the U.S.A. and Hancock (1843) in the U.K. Both found that when natural rubber was heated with sulphur, the undesirable properties of surface tackiness and creep under stress could be eliminated. The chemical reaction involves the formation of interchain links, composed of two, three, or four sulphur atoms, between sites of unsaturation on adjacent chains. It has been found that about three parts sulphur per hundred parts rubber produces a useful elastomer, capable of reversible extensions of up to 700 per cent. Increasing the sulphur, up to 30 parts per hundred, alters the material drastically and produces a hard, highly crosslinked substance called Ebonite. The actual mechanism of the cross-linking reaction is still in some doubt but it is thought to proceed via an ionic route.

The quantitative introduction of crosslinks using sulphur is difficult to achieve. An alternative method involves heating the polymer with either dicumyl peroxide or ditertiary butyl peroxide and this is applicable to both polydienes and elastomers with no unsaturated sites in the chain, *e.g.* the ethylene-propylene copolymers or the polysiloxanes. The peroxide radical abstracts hydrogen from the polymer chain and creates a radical site on the interior of the chain. Two such sites interact to form the crosslink. It is claimed that one peroxide molecule produces one crosslink, but side reactions may impair the attainment of such a precise ratio. The major disadvantage of this technique is its commercial inefficiency; better results can be obtained by synthesizing precursors which are more suitable for crosslinking, such as in ethylene-propylene terpolymers. Recently room temperature vulcanization techniques have been developed for silicone elastomers. These are based on linear polydimethylsiloxane chains terminated by hydroxyl groups. Curing can be achieved either by adding a crosslinking agent and a metallic salt catalyst such as tri- or tetra-alkoxysilane with stannous octoate or by incorporating in the mixture a crosslinking agent sensitive to atmospheric water which initiates vulcanization.

A crosslinking technique, recently developed, makes use of the reactive nitrene intermediates formed from compounds of the type $N_3COO(CH_2)_nOOCN_3$. If these are heated in the presence of linear polymers such as polyethylene or polypropylene, nitrogen is lost and the resulting dinitrene reacts with the polymer chains to crosslink them. This reaction is particularly useful when no unsaturated sites exist in the chain. The crosslinking also serves to improve the resilience of polyethylene and other related polymers.

13.3 Properties of elastomers

Elastomers exhibit several other unusual properties which can be attributed to their chain-like structure. It has been found that (a) as the temperature of an elastomer increases, so too does the elastic modulus, (b) an elastomer becomes warm when stretched, and (c) the expansivity is positive for an unstretched sample but negative for a sample under tension. As these properties are so different from those observed for other materials they are worth examining in detail.

In simple mechanistic terms, the elastic modulus is simply a measure of the resistance to the uncoiling of randomly oriented chains in an elastomer sample under stress. Application of a stress eventually tends to untangle the chains and align them in the direction of the stress, but an increase in temperature will increase the thermal motion of the chains and make it harder to induce orientation. This leads to a higher elastic modulus. Under a constant force some chain orientation will take place, but an increase in temperature will stimulate a reversion to a randomly coiled conformation and the elastomer will contract.

While this is a satisfactory picture to describe properties (a) and (c), it is possible to derive the more rigorous thermodynamic explanation which follows.

13.4 Thermodynamic aspects of rubber-like elasticity

As early as 1806 John Gough made two interesting discoveries when studying natural rubber. He found that (1) the temperature of rubber changed when a rapid change in sample length was induced, and (2) a rubber sample under constant tension changed length as the temperature changed, *i.e.* observations (b) and (c) above.

The first point is readily demonstrated with a rubber band. If the centre of the band is placed lightly touching the lips and then extended rapidly by pulling on both ends, a sensation of warmth can be felt as the temperature of the rubber rises.

In thermodynamic terms process (1) is analogous to the change in temperature undergone by a gas subjected to a rapid volume change and can be treated formally in a like manner. An ideal gas can only store energy in the form of kinetic energy and when work is performed on a gas during compression, the energy appears as kinetic energy or heat causing the temperature to rise. Extension of an elastomer results in the evolution of heat for similar reasons.

This effect can be examined further by studying the *reversible adiabatic extension* of an elastomer. Although this experiment is more easily carried out under conditions of constant pressure rather than constant volume, it is best to derive the relevant equations for constant volume. For this reason we consider first the Helmholtz function for the system

$$A = U - TS. \tag{13.1}$$

If the applied force is f, and l° and l are the lengths of the sample in the unextended and extended states then differentiation of equation (13.1) with respect to l, at constant temperature gives

$$(\partial A/\partial l)_T = (\partial U/\partial l)_T - T(\partial S/\partial l)_T \tag{13.2}$$

The work done by the system during a reversible extension of the sample by an amount dl against the restoring force f, is given by $dA = -f dl$ and so

$$f = (\partial U/\partial l)_T - T(\partial S/\partial l)_T = f_U + f_S \tag{13.3}$$

The force f is seen to be composed of two contributions; the energy f_U and the entropy f_S. For an ideal elastomer the contribution of f_U to the total force is negligible because there is no energy change during extension and

$$f = -T(\partial S/\partial l)_T. \tag{13.4}$$

This is the expression for an entropy spring and shows that the strain in a stretched elastomer is caused by a reduction in conformational entropy (see section 7.1) of the chains under stress. If the interdependence of length and temperature is now examined, the quantity of interest is $(\partial T/\partial l)$ and

$$(\partial T/\partial l)_{S, p} = - (\partial T/\partial S)_{l, p}(\partial S/\partial l)_{T, p}. \tag{13.5}$$

Each factor can now be evaluated separately as follows

$$(\partial T/\partial S)_{l, p} = (\partial T/\partial H)_{l, p} (\partial H/\partial S)_{l, p} = T/C_{p, l}, \tag{13.6}$$

while starting from the Maxwell equation $(\partial A/\partial T)_V = -S$ and differentiating both sides with respect to l, leads to

$$(\partial/\partial l)(\partial A/\partial T) = (\partial/\partial T)(\partial A/\partial l) = (\partial f/\partial T)_l = -(\partial S/\partial l)_T. \tag{13.7}$$

It follows then from equations (13.5), (13.6), and (13.7) that

$$(\partial T/\partial l)_{S, p} = (T/C_{p, l})(\partial f/\partial T)_{l, p}. \tag{13.8}$$

This shows that for a rapid reversible adiabatic extension when $(df/dT)_{l, p}$ is positive, the temperature of the elastomer increases. The equation also tells us that the elastomer will contract, not expand on heating.

Demonstration of the Gough effect. This thermoelastic effect can be conveniently demonstrated using an apparatus like that shown in figure 13.1(a). A spring with one end free and the other attached to a frame is hooked onto a pointer P, pivoted on the support rod R. A rubber band is stretched between a fixed hook X and the other end Y of the spring-pointer arrangement. If the band is encased in a glass tube it can be heated with a bunsen producing a contraction which is indicated by the movement of the pointer. Results showing the temperature dependent behaviour of an elastomer held under constant tension, are plotted in figure (13.1b). Between 210 and 330 K the tensile force increases as T increases indicating a contraction of the elastomer, but the behaviour is reversed below 210 K as the material passes through the glass transition and the polymer reacts normally like a glassy solid.

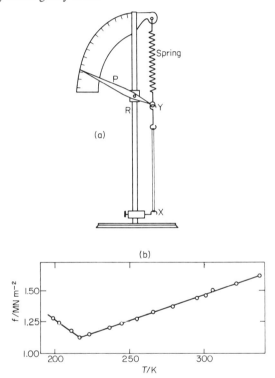

FIGURE 13.1. (a) Apparatus used to demonstrate thermoelastic effects in elastomers. (b) Tensile force f — temperature plot for rubber. The minimum occurs in the region of the glass transition temperature. (From data of Meyer and Ferri, 1935.)

13.5 Non-ideal elastomers

The behaviour of most elastomers under stress is far from ideal and a significant contribution from f_U is found. This can be expressed in several ways and following on from equations (13.3) and (13.7):

$$f_S = T(\partial f/\partial T)_{V, l},$$ (13.9)

so that

$$(f_U/f) = (1 - f_S/f) = 1 - (T/f)(\partial f/\partial T)_{V, l},$$ (13.10)

or $\quad (f_U/f) = -\{\partial \ln(f/T)/\partial \ln T\}_{V, l}.$ (13.11)

The experimental determination of the quantity of $(\partial f/\partial T)_{V, l}$ at constant volume is extremely difficult to perform and it is much more convenient to work at constant pressure. Approximate relations between the quantities have been suggested by Flory who proposed that equation (13.11) be modified by an additional term

$$f_u/f = -\{\partial \ln(f/T)/\partial \ln T\}_{p,l} - \alpha T/\lambda^3 - 1)$$ (13.12)

where λ is the extension ratio (l/l°) and α is the expansivity. The modifying factor is the difference between the values of $(\partial f/\partial T)$ measured at constant volume and constant pressure.

Thermoelastic data showing the relative contributions of f_U and f_S appear as the curves in figure 13.2 and it can be seen that at low extensions f_U remains small but begins to increase as the elongation rises.

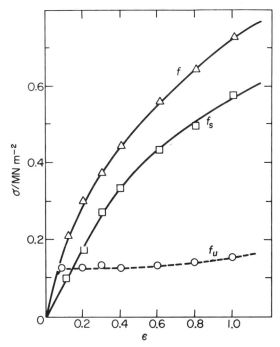

FIGURE 13.2. Stress-strain $(\sigma - \epsilon)$ curves derived from data in figure 13.3, showing the magnitude of the contributions from f_U and f_S. (Adapted from Beevers *Experiments in Fibre Physics*.)

Experimental determination of f_U and f_S. The curves in figure 13.2 can be derived using a stress-relaxation balance. (*cf.* chapter 12.) A rubber band is stretched between two hooks and the force required to maintain a constant deformation is measured at various temperatures. If a number of extensions are examined, a family of curves (figure 13.3) can be constructed starting with the highest temperature at each strain. As the temperature is reduced incrementally, the load must be adjusted accordingly. Here the stress is taken as the ratio of the load to the unstrained cross-sectional area of the elastomer (best measured simply with an accurate ruler), while the strain is just the extension of the sample.

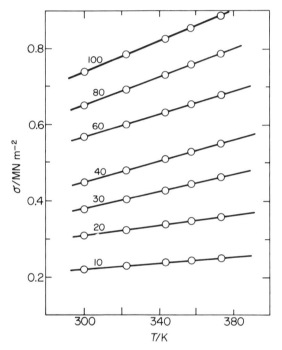

FIGURE 13.3. Thermoelastic behaviour of a rubber sample. Stress-temperature $(\sigma - T)$ curves for a series of extension values. The percentage strain is shown against each curve. (Adapted from Beevers.)

Values of f_S and f_U at each strain are calculated from the slope and the intercept at 0 K respectively. In the diagram $f_S = -(T\partial S/\partial l)$ has been calculated for $T = 298$ K.

13.6 Distribution function for polymer conformation
The retractive force in a rubbery material is a direct result of the chain in the extended form wanting to regain its most probable, highly coiled conformation. Thus it is of considerable interest to calculate, in addition to the average

dimensions of the polymer chain, the distribution of all the possible shapes available to the molecules experiencing thermal vibrations.

This can be accomplished by first considering a chain in three-dimensional space with one end located at the origin of a set of Cartesian co-ordinates. The probability that the other end will be found in a volume element (dx, dy, dz) at the point (x, y, z) is given by $p(x,y,z)dx,dy,dz$, where $p(x,y,z)$ measures the number of possible conformations the chain can adopt in the range $(x + dx)$, $(y + dy)$, and $(z + dz)$. This is known as the *probability density* and can be expressed in terms of the parameter $\beta = (3/2nl^2)^{1/2}$ as

$$(dx,dy,dz)p(x,y,z) = (\beta^3/\pi^{3/2}) \exp\{-\beta^2(x^2 + y^2 + z^2)\} dx.dy.dz. \qquad (13.13)$$

It now remains to calculate the probability $p(r)$ that the chain is located in a spherical shell of thickness dr and distance r from the origin. This is given by

$$p(r)dr = 4\pi r^2 \, dr \{(\beta^3/\pi^{3/2}) \exp(-\beta^2 r^2)\}, \qquad (13.14)$$

where $4\pi r^2 \, dr$ is the volume of the shell. This function has the form shown in figure 13.4 where the maximum corresponds to the most probable distance between chain ends and is given by

$$r^2 = \int_0^{\infty} r^2 p(r)dr = 3/2\beta^2 = l^2 n. \qquad (13.15)$$

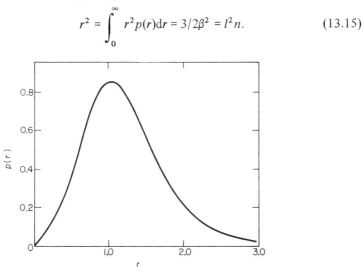

FIGURE 13.4. Distribution function $p(r)$ for end-to-end distances r calculated from equation (13.14). The maximum occurs at $r = 1/\beta$.

It has already been stated that while a highly coiled conformation is the most probable for an elastomer the energy of an ideal elastomer in the extended form is the same as that in the coiled state. The elastic retractive force is entropic and not energetic in origin and depends on the fact that the number of possible ways a polymer coil can exist in a highly compact form is overwhelmingly greater than the number of available arrangements of the chain segments in an extended ordered form. This means that the probability of finding a chain in a coiled state is high and as probability and entropy are related by the Boltzmann equation,

$$S = k \ln p, \tag{13.16}$$

then the most likely state for the chain is one of maximum entropy. Substitution in equation (13.14) leads to

$$S = C - k\beta^2 r^2, \tag{13.17}$$

where C is a constant. This relation provides a measure of the entropy of an ideal flexible chain whose ends are held a distance r apart.

While an elastomer is considered to be solely an entropy spring, it is never perfect and small changes in the internal energy are observed when it is under stress.

13.7 Statistical approach

Having examined the thermodynamic approach we can now outline briefly the stress–strain behaviour of an elastomer in terms of the chain conformations.

Consider a lightly crosslinked network with the junction points sufficiently well spaced to ensure that the freedom of movement of each chain section is unrestricted. If the length of a chain between two crosslinking points is assumed to be r, the probability distribution calculated above can be applied to the network structure. The entropy of a single chain, as described by equation (13.17) can be used to calculate S for a chain in the network, but if the stress-strain relations are required, then S for chains in both the deformed and unde-formed states must be calculated, and for the complete network an integration over all the chains in the sample should be performed.

When a unit cube of elastomer is stretched, the resulting entropy change is

$$\Delta S = -\tfrac{1}{2}Nk(\lambda_1^2 + \lambda_2^2 + \lambda_3^2 - 3), \tag{13.18}$$

where N is the number of individual chain segments between successive crosslinks per unit volume, and λ_1, λ_2, and λ_3 are the principal extension ratios. For an ideal elastomer there is no change in the internal energy and the work of deformation w is derived entirely from $w = -T\Delta S$ so that

$$w = \tfrac{1}{2}NkT(\lambda_1^2 + \lambda_2^2 + \lambda_3^2 - 3). \tag{13.19}$$

EXPERIMENTAL STRESS-STRAIN RESULTS

The treatment of mechanical deformation in elastomers is simplified when it is realized that the Poisson ratio (table 12.1) is almost 0.5. This means that the volume of an elastomer remains constant when deformed and if one also assumes that it is essentially incompressible ($\lambda_1\lambda_2\lambda_3 = 1$) the stress-strain relations can be derived for simple extension and compression using the stored energy function w.
Simple extension. The required conditions are $\lambda = \lambda_1$ and $\lambda_2 = \lambda_3 = \lambda^{-1/2}$.

Elimination of λ_3 and substitution of λ leads to

$$w = \tfrac{1}{2}NkT(\lambda^2 + 2/\lambda - 3). \tag{13.20}$$

As the force f is simply $= (dw/d\lambda)$, differentiation provides a relation between f and λ

$$f = G(\lambda - \lambda^{-2}), \tag{13.21}$$

where the modulus factor is $G = NkT$ and shows that G will depend on the number of crosslinks in the sample. If we neglect the fact that the chains in a network possess free ends and assume that all network chains end at two cross-linking points then

$$G = RT\rho M_s^{-1}, \tag{13.22}$$

where R is the gas constant, ρ the polymer density, and M_s is the number average molar mass of a chain section between two junction points in the network. The product (ρM_s^{-1}) is then a measure of the crosslink density of a sample. This also shows that the modulus will increase with temperature. The statistically derived equation (13.21) was tested first by Treloar for both extension and compression of a rubber vulcanizate. A value of $G = 0.392$ MN m^{-2} was chosen to fit the data at low extensions where the application of Gaussian statistics to the chain might be expected to be valid. The experimental data for extension, figure 13.5 drops below the theoretical curve at $\lambda \approx 2$ then sweeps up sharply when λ exceeds 6.

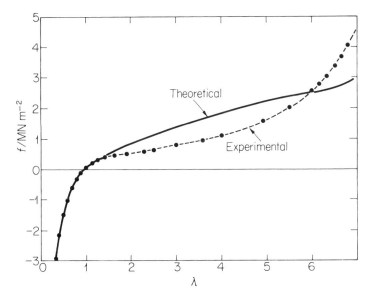

FIGURE 13.5. Extension or compression ratio λ as a funticon of the tensile or compressive force f for a rubber vulcanizate. Theoretical curve is derived from equation (13.21) using $G = 0.392$ MNm^{-2}. (From data by Treloar, 1944.)

This rapid increase was originally attributed to the crystallization of the polymer under tension but it is now believed to reflect the departure of the network from the assumed Gaussian distribution.

Simple compression. The agreement between experiment and theory for the equivalent of simple compression is somewhat better. The experimental determination of compression is rather difficult to carry out, but a two-dimensional extension with $\lambda_2 = \lambda_3$ serves to provide the same information. This is achieved by clamping a circular sheet of rubber around the circumference and then inflating it to produce a stress. Correlation between the results and a curve derived from equation (13.21), using the same G factor as before, is good.

The statistical approach, using a Gaussian distribution, thus appears to predict the stress-strain response except at moderately high elongations.

Pure shear. The corresponding equation to describe the behaviour of an elastomer under shear is $f = G(\lambda - \lambda^{-3})$ but again agreement is reasonable only at low λ.

Large elastic deformation. At high extensions, departure from the Gaussian chain approximation becomes significant, and has led to the development of a more general, but semi-empirical, theory based on experimental observations. This is expressed in the Mooney, Rivlin, and Saunders, MRS equation

$$\tfrac{1}{2}f(\lambda - \lambda^{-2})^{-1} = C_1 + C_2\lambda^{-1}, \tag{13.23}$$

where C_1 and C_2 are constants. Unfortunately this simple form is still only capable of predicting data over a range of low to moderate extensions, but not for samples under compression.

13.8 Swelling of elastomeric networks

A crosslinked elastomer cannot dissolve in a solvent. Dispersion is resisted because the crosslinks restrict the movement and complete separation of the chains, but the elastomer does swell when the solvent molecules diffuse into the network and cause the chains to expand. This expansion is counteracted by the tendency for the chains to coil up and eventually an equilibrium degree of swelling is established which depends on the solvent and the crosslink density, *i.e.* the higher the crosslink density the lower the swelling.

The behaviour is predicted in the Flory-Huggins treatment of swelling which leads to a relation between the degree of swelling Q, for a particular solvent and the shear modulus G of the unswollen rubber

$$G = RTA/V_1 Q^{5/3}, \tag{13.24}$$

where A is a constant. The validity of this expression was tested by Flory as shown in figure 13.6, where the expected slope of $-5/3$ was obtained. The theory also predicts that the equilibrium swelling of an elastomer increases when under a tensile stress.

The statistical theory will also describe the response under stress of elastomers swollen by solvents and in general it is found that the greater the degree of

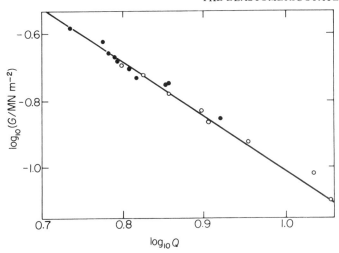

FIGURE 13.6. The swelling of butyl vulcanizates by cyclohexane plotted as a function of the elastic modulus, using equation (13.23). (Flory, 1946.)

swelling the better the agreement between theory and experiment. The modifications necessary for the treatment of swollen networks are relatively straightforward and the stored energy function becomes

$$w = \tfrac{1}{2} N k T \phi_r^{1/3} (\lambda_1'^2 + \lambda_2'^2 + \lambda_3'^2 - 3), \tag{13.25}$$

where ϕ_r is the volume fraction of the elastomer and the prime represents the swollen unstrained state of the network. Similarly for simple extension $\phi_r^{1/3}$ is inserted in the right-hand side of equation (13.21).

13.9 Network defects

The number average molar mass M_s of a chain section between two junction points in the network is an important factor controlling elastomeric behaviour; when M_s is small the network is rigid and exhibits limited swelling, but when M_s is large the network is more elastic and swells rapidly when in contact with a compatible liquid. Values of M_s can be estimated from the extent of swelling of a network, which is considered to be ideal but rarely is, and interpretation of the data is complicated by the presence of network imperfections. A real elastomer is never composed of chains linked solely at tetra-functional junction points, but will inevitably contain defects such as (a) loose chain ends, (b) intramolecular chain loops, and (c) entangled chain loops.

For a swollen network at equilibrium we have

$$\ln(1 - \phi_r) + \phi_r + \chi_1 \phi_r^2 = - V_1 \phi_r^{1/3} \rho \{ M_s^{-1} g'(1 - 2M_s M_n^{-1}) \}, \tag{13.26}$$

where ϕ_r is the volume fraction of the elastomer, g' is an empirical constant (≈ 3) and M_n is the molar mass prior to crosslinking. The correction factor in the brackets is believed to arise mainly from the presence of free chain ends.

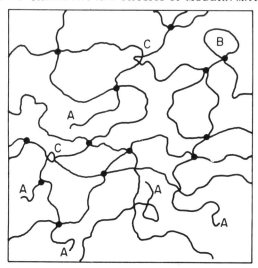

FIGURE 13.7. Diagram showing defects in an elastomeric network; A, loose chain ends; B, intramolecular chain loops; C, entangled chain loops.

Resilience of elastomers

When an elastomer, in the form of a ball, is dropped from a given height onto a hard surface, the extent of the rebound provides an indication of the resilience of the elastomer. A set of elastomer balls, manufactured by Polysar Corporation in Canada, provide an excellent demonstration of this phenomenon and have been used to measure the rebound height of several elastomers, recorded schematically in figure 13.8. If h_0 is the original height and h is the recovery height, then the rebound resilience is defined as (h/h_0) and the relative energy loss per half cycle is $(1 - h/h_0)$.

It should be remembered that an elastomer exhibiting good elastic properties under slow deformation may not possess good resilience; the relative recoveries of natural and butyl rubber provide a good example of this fact. As resilience is the ability of an elastomer to store and return energy when subjected to a rapid deformation, it can be shown that temperature also plays an important part in determining resilience. If the butyl and natural rubber balls are now heated to about 373 K, both will rebound to about the same extent. The importance of the two variables, time and temperature, is illustrated in a plot of the rebound resilience against temperature for three elastomers and a recognized plastic, figure 13.9. The sharply defined minima are characteristic of such curves and the broad butyl curve is anomalous. The minimum for curve 1 is closely related to the loss of long range elasticity at the glass temperature $T_g = 218$ K although it actually occurs at the higher temperature of 238 K. A similar situation is found for neoprene and poly(methyl methacrylate). One can conclude from this that resilience is closely related to the molecular structure and the intermolecular forces affecting the ability of the chain to rotate.

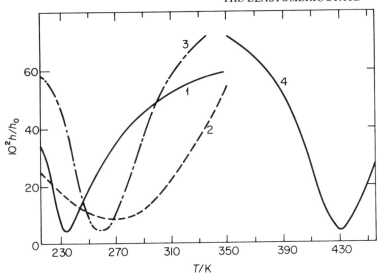

FIGURE 13.8. Change of rebound resilience (h/h_0) with temperature T for: 1, natural rubber; 2, butyl rubber; 3, neoprene; 4, poly(methyl methacrylate). (After Mullins (1947) and Gordon (1957).)

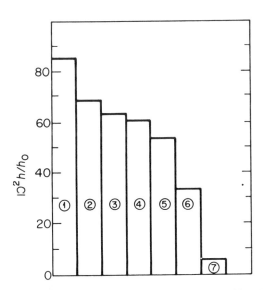

FIGURE 13.9. Percentage rebound recovery measured at 298 K, with balls made from: 1, *cis*-polybutadiene; (2), synthetic *cis*-polyisoprene; (3), natural *cis*-polyisoprene; (4), ethylene-propylene copolymer; (5), SBR; (6), *trans*-polyisoprene; (7), butyl rubber.

When the chains are deformed during a bounce, a stress is applied and then rapidly removed. The time required for the chains to regain their original positions is measured by the relaxation time τ, defined in section 12.4. Thus relaxation times are a measure of the ability of the chains to rotate. At room temperature the butyl rubber with the bulky methyl groups will not rotate as readily as the cis-polyisoprene so that when the deformation of chains in the sample of butyl rubber occurs, the chains do not return to their equilibrium positions as rapidly as the natural rubber, i.e. τ is longer.

The elastomer showing superior rebound potential at room temperature is cis-polybutadiene. This sample is non-crystalline and has no pendant groups to impede free segmental rotation, so the relaxation time is correspondingly shorter than other elastomers. The response of the butyl sample improves as the temperature increases because additional thermal energy is available to enhance chain rotation and decrease the relaxation time correspondingly. This leads to an improved resilience and the rebound potential now matches the natural rubber whose τ is not so sensitive to temperature change in this range.

General Reading

R. B. Beevers, *Experiments in Fibre Physics.* Butterworths (1970).
F. Bueche, *Physical Properties of Polymers,* Chapter 1. Interscience Publishers Inc. (1962).
P. J. Flory, *Principles of Polymer Chemistry,* Chapter 11. Cornell Univ. Press, Ithaca, N.Y. (1953).
R. Meares, *Polymers: Structure and Bulk Properties,* Chapters 6–8, Van Nostrand (1965).
D. A. Smith, *Addition Polymers,* Chapter 10. Butterworths (1968).
A. V. Tobolsky and H. Mark, *Polymer Science and Materials,* Chapter 9. Wiley-Interscience (1971).
L. R. G. Treloar, *Physics of Rubber Elasticity.* Clarendon Press (1958).

References

1. P. J. Flory, *Ind. Eng. Chem.,* **38**, 417 (1946).
2. K. H. Meyer and C. Ferri, *Helv. Chim. Acta,* **18**, 570 (1935).
3. L. Mullins, *I.R.I. Trans.,* **22**, 235 (1947).
4. L. R. G. Treloar, *Trans. Farad. Soc.,* **40**, 59 (1944).

CHAPTER 14

Structure–Property Relations

14.1 General considerations

The increasing use of synthetic polymers by industrialists and engineers to replace or supplement more traditional materials, such as wood, metals, ceramics, and natural fibres, has stimulated the search for even more versatile polymeric structures covering a wide range of properties. For such a quest to be efficient, a fundamental knowledge of structure–property relations is required.

The problem can be examined initially on two broad planes:

(a) *The chemical level.* This deals with information on the fine structure, namely what type of monomer constitutes the chain and whether more than one type of monomer is used (copolymer), *i.e.* the parameters which relate ultimately to the three-dimensional aggregated structure, and influence the extent of sample crystallinity and the physical properties.

(b) *The architectural aspects.* These are concerned with the chain as a whole, and now we are required to ask such questions as: is the polymer linear, branched, or crosslinked; what distribution of chain lengths exist; what is the chain conformation and rigidity?

Having considered these general points, one must then establish the suitability of a polymer for a particular purpose. This depends on whether it is glass-like, rubber-like, or fibre forming, and the characteristics depend primarily on chain flexibility, chain symmetry, intermolecular attractions, and of course environmental conditions. Excluding the environment, these parameters, in turn, are reflected in the more tangible factors, T_m, T_g, modulus, and crystallinity, which, being easier to assess, are commonly used to characterize the polymer and ascertain its potential use.

As the relative values of both T_m and T_g play such an important part in determining the ultimate behaviour of a polymer, we can begin an examination of structure and properties by finding out how a polymer scientist can attempt to control these parameters.

14.2 Control of T_m and T_g

We have already seen, in earlier chapters, how chain symmetry, flexibility, and tacticity, can influence the individual values of both T_m and T_g. Thus a highly flexible chain has a low T_g, which increases as the rigidity of the chain becomes greater. Similarly, strong intermolecular forces tend to raise T_g and also increase crystallinity. Steric factors play an important role. A high T_g is obtained when large pendant groups attached to the chain restrict its internal rotation, and bulky pendant groups tend to impede crystallization, except when arranged regularly in isotactic or syndiotactic chains.

Chain flexibility is undoubtedly the controlling factor in determining T_g, but it also has a strong influence on T_m. Hence we must consider these parameters together from now on and determine how to effect control of both.

CHAIN STIFFNESS

It is important to be able to regulate the degree of chain stiffness, as rigid chains are preferred for fibre formation, while flexible chains make better elastomers. The flexibility of a polymer depends on the ease with which the backbone chain bonds can rotate. Highly flexible chains will be able to rotate easily into the various available conformations, while the internal rotations of bonds in a stiff chain are hindered and impeded.

Variations in chain stiffness can be brought about by incorporating different groups in linear chains and the results can be appraised by following the changes in T_m and T_g in a series of different polymers. The effects can be assessed more easily if an arbitrary reference is chosen and the simplest synthetic organic polymer, polyethylene with $T_m \approx 400$ K and $T_g \approx 188$ K, is suitable for this purpose.

One can begin by considering a general structure $-[(CH_2)_{\overline{m}}-X]_n$, where m and X vary. The effect on T_m of incorporating different links in the carbon chain is illustrated in table 14.1.

TABLE 14.1. Influence of various links on T_m when incorporated in an all-carbon chain

Polymer group	Repeat unit	T_m/K					
		m	2	3	4	5	6
Polyethylene	$-[(CH_2)_m]-$		400	–	–	–	–
Polyester	$-[(CH_2)_m CO \cdot O]-$		395	335	329	335	325
Polycarbonate	$-[(CH_2)_m - O \cdot CO \cdot O]-$		312	320	330	318	320
Polyether	$-[(CH_2)_m \cdot CH_2 - O]-$		308	333	–	–	–
Polyamide	$-[(CH_2)_m - CO \cdot NH]-$		598	538	532	496	506
Polysulphone	$-[(CH_2)_m CH_2 \cdot SO_2]-$		573	544	516	493	–

The chain flexibility is increased by groups such as $-\!\!(\!\!-O\!\!-\!\!)\!\!-$, $-\!\!(\!\!-CO.O\!\!-\!\!)\!\!-$, and $-\!\!(\!\!-OCO.O\!\!-\!\!)\!\!-$, and as the length of the $-\!\!(\!\!-CH_2\!\!-\!\!)\!\!-$ section grows. This is shown by a lowering of T_m relative to polyethylene. Insertion of the polar $-\!\!(\!\!-SO_2\!\!-\!\!)\!\!-$ and $-\!\!(\!\!-CONH\!\!-\!\!)\!\!-$ groups raises T_m, because the intermolecular bonding now assists in stabilizing the extended forms in the crystallites.

Chain stiffness is also greatly increased when a ring is incorporated in the chain, as this restricts the rotation in the backbone and reduces the number of conformations a polymer can adopt. This is an important aspect, as fibre properties are enhanced by stiffening the chain, and the effect of aromatic rings on T_g and T_m is shown in table 14.2.

TABLE 14.2. Effect of aromatic rings on chain stiffness, as shown by the values of T_m and T_g

Structure	T_g/K	T_m/K
1. $-\!\!(\!\!-CH_2\!\!-\!\!CH_2\!\!-\!\!)_{\overline{n}}$	188	400
2. $-\!\!(\!\!-CH_2\!\!-\!\!CH_2\!\!-\!\!O\!\!-\!\!)_{\overline{n}}$	206	339
3. $-\!\!(\!\!-CH_2\!\!-\!\!\bigcirc\!\!-\!\!CH_2\!\!-\!\!)_{\overline{n}}$	–	about 653
4. $\left[-\!\!(\!\!-CH_2)_2\!\!-\!\!O\!\cdot\!CO\!\!-\!\!\bigcirc\!\!-\!\!CO.O\!\!-\!\!\right]_n$	342	538
5. $-\!\!(\!\!-NH(CH_2)_6NHCO.(CH_2)_4CO\!\!-\!\!)_{\overline{n}}$	320	538
6. $\left[-NH\!\!-\!\!\bigcirc\!\!-\!\!NHCO(CH_2)_4\!\!-\!\!CO\!\!-\!\!\right]_n$	–	613
7. $\left[-NH\diagdown\!\!\bigcirc\!\!\diagup NHCO\diagdown\!\!\bigcirc\!\!\diagup CO\!\!-\!\!\right]_n$	546	about 635 (Decomposition)
8. $\left[-NH\!\!-\!\!\bigcirc\!\!-\!\!NHCO\!\!-\!\!\bigcirc\!\!-\!\!CO\!\!-\!\!\right]_n$	–	about 773

The p-phenylene group in structure 3 causes a big increase in T_m and this can be modified by introducing a flexible group as in terylene, structure 4. This shows that a judicious combination of units can lead to a wide variety of chain flexibilities and physical properties. The effect of the aromatic ring is again obvious in structures 5 through 8. The influence of chain symmetry is also noticeable when comparing the decomposition temperature of the symmetrical chain 8 with the unsymmetrical chain 7. The latter is also an inefficient close-packing structure, as sufficient disorder exists for the glass transition to appear.

INTERMOLECULAR BONDING

An increase in the lattice energy of a crystallite is obtained when the three-dimensional order is stabilized by intermolecular bonding. In the polyamide and polyurethane series, the additional cohesive energy of the hydrogen bond (about 24 kJ mol^{-1}) strengthens the crystalline regions and raises T_m. The effect is strongest when regular, evenly spaced, groups exist in the chain, as with nylon-6,6. The importance of secondary bonding in the polyamide series is illustrated quite dramatically when the crucial hydrogen atom of the amide group is replaced by a methylol. The loss of the hydrogen-bonding capability impairs the tendency for regular chain alignment to take place and the character of the polyamide changes dramatically. With little substitution, they are suitable fibres, but, as the hydrogen is replaced, they change and become more elastomeric, then eventually like balsams, and finally liquids.

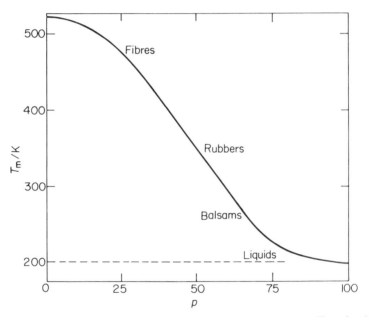

FIGURE 14.1. Change in the properties and melting temperature T_m of nylon-6,6 as the hydrogen-bonding capacity is reduced by changing the percentage p of amide substitution. (Adapted from R. Hill, *Fibres from Synthetic Polymers.*)

An alternative method of reducing the hydrogen-bonding potential, and so T_m, in the polyamides, is to increase the length of the $-(\text{CH}_2-)_{\overline{n}}$ sequence between each bonding site. This leads to nylons with a variety of properties, e.g. nylon-12 has properties intermediate between those of nylon-6 and polyethylene. The effect is shown in table 14.3.

14.3 Relation between T_m and T_g

Most of the factors discussed so far influence T_g and T_m in much the same way, but in spite of this, the fact that T_m is a first-order thermodynamic transition, whereas T_g is not, precludes the possibility of a simple relation between them.

TABLE 14.3. Melting temperatures T_m of linear aliphatic polyamides

Monadic nylon	T_m/K	Dyadic nylon	T_m/K
4	533	4,6	581
6	496	5,6	496
7	506	6,6	538
8	473	4,10	509
9	482	5,10	459
10	461	6,10	495
11	463	6,12	482
12	452		

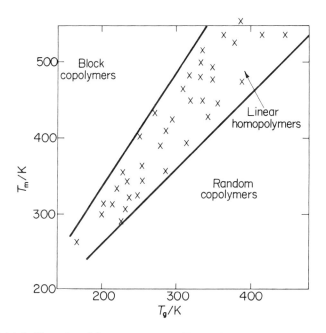

FIGURE 14.2. Plot of melting temperature T_m against glass transition temperature T_g for linear homopolymers with (T_g/T_m) lying in the range 0.5 to 0.8.

There is, however, a crude correlation, represented in figure 14.2, where a broad band covers most of the results for linear homopolymers, and the ratio (T_g/T_m) lies between 0.5 and 0.8 for about 80 per cent of these.

Obviously then, a synthetic chemist attempting to control the T_m and T_g of a simple chain structure by varying flexibility, symmetry, tacticity, *etc.*, is limited to structures with either a high T_m and T_g or a low T_m and T_g. In effect neither T_m nor T_g can be controlled separately to any great degree.

To exercise this additional control another method of chain modification must be sought and this leads to the use of copolymers.

14.4 Random copolymers

Axial symmetry in a chain is a major factor in determining the ability of a chain to form crystallites, and one method of altering the crystalline content is to incorporate some structural irregularity in the chain. The controlled inclusion of linear symmetrical homopolymer chains $+ A +_n$ in a crystal lattice can be achieved by copolymerizing A with varying quantities of monomer B, whose purpose is to destroy the regularity of the structure.

This leads to a gradual decrease in T_m as shown schematically in figure 14.3. The broken line represents the possibility that, in the middle composition range, the decrease in regularity is so great that the material is amorphous. This situation is sometimes obtained when a terpolymer is prepared.

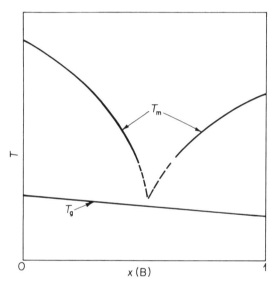

FIGURE 14.3. Schematic representation of T_m and T_g plotted as functions of copolymer composition shown as mole fraction $x(B)$ of B. The broken lines represent the possibility that structural irregularities are so great that no crystallization of the copolymer can occur.

A practical application is found in the polyamides. An improvement in the elastic qualities of the polyamide fibre is obtained if the modulus is reduced and, as the factors which affect the melting temperature affect the modulus, this can be achieved by starting from nylon-6,6 or nylon-6,10 and forming (66/610) copolymers. The random inclusion of the two types of unit in the chain disturbs both the symmetry and the regular spacing of the hydrogen-bonding sites, resulting in a drop in T_m.

The glass transition T_g is not affected in the same way as T_m, because T_g is more a function of the differences in chain flexibility, than the packing efficiency.

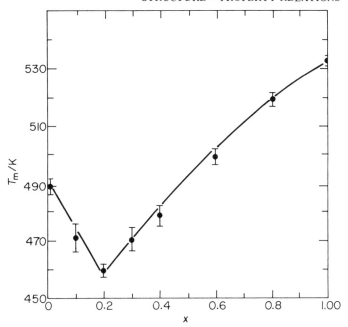

FIGURE 14.4. Melting temperatures of random copolymers of nylon-6,6 and nylon-6,10 as a function of the mole fraction of adipamide in the copolymer. (From data by Cowie and Mudie.)

This means that the response of T_g to a change in copolymer composition is quite different and this affords a means of controlling the magnitudes of T_m and T_g independently, a most important feature not readily achieved by other methods.

14.5 Dependence of T_m and T_g on copolymer composition

A quantitative expression for the depression of the melting temperature can be derived, thermodynamically, in terms of the composition and enthalpy of fusion ΔH_u of polymer A by

$$1/T_m^{AB} \quad 1/T_m^{A} -- (R/\Delta H_u) \ln x_A, \qquad (14.1)$$

where T_m^A and T_m^{AB} are the melting temperatures of pure polymer A and the copolymer AB respectively, and x_A is the mole fraction of A in the copolymer.

The simple linear relation between T_g and x shown in figure 14.3 is found only for a few copolymers composed of compatible monomer pairs, such as styrene copolymerized with either methyl acrylate or butadiene. A simple ideal mixing rule can be applied to these systems, but when the comonomer properties differ markedly, the linear dependence is lost, and a non-linear equation has to be developed.

One simple relation, which usefully describes the behaviour of many vinyl monomer pairs is

$$1/T_g^{AB} = w_A/T_g^A + w_B/T_g^B, \qquad (14.2)$$

where w_A and w_B are the mass fractions of monomers A and B. For a system where the conditions $T_g^A < T_g^{AB} < T_g^B$ hold, the free volume concept can be used to formulate a relation between T_g and w, and Gordon and Taylor have proposed

$$(T_g^{AB} - T_g^A) w_A + K(T_g^{AB} - T_g^B) w_B = 0. \qquad (14.3)$$

This expression assumes that the free volume contribution from a monomer is the same as both homo- and copolymers. For a given pair of monomers the constant K is calculated from the corresponding expansivities of the homopolymers.

$$K = (\alpha_l^B - \alpha_g^B)/(\alpha_l^A - \alpha_g^A). \qquad (14.4)$$

A similar relation has been proposed by Gibbs and Di Marzio,

$$(T_g^{AB} - T_g^A)n_0^A + (T_g^{AB} - T_g^B)n_0^B = 0, \qquad (14.5)$$

where now the fraction of rotatable bonds n_0 is introduced in place of the composition term.

14.6 Block copolymers

Random copolymers can be prepared if one wishes to narrow the gap between T_m and T_g in a sample, and so cover one property region not readily satisfied by homopolymers.

If a wider interval between T_m and T_g is required, a different class of co-polymer — the block copolymer — must be investigated. These are usually $\{AB\}$ or $\{ABA\}$ block sequences. By synthesizing sequences, which are long enough to crystallize independently, the combination of a high melting block A with a low melting block B will provide a material with a high T_m from A and a low T_g from B. A slight depression of T_m, arising from the presence of block B, is sometimes encountered, but this is rarely large. Combinations of this type allow the scientist to cover the remaining area of property combinations shown in figure 14.2.

The change of T_g in block copolymers is rather variable and certain pairs of monomers will form a block copolymer possessing two glass transitions. Interesting changes in the mechanical properties can be obtained when SBR block copolymers are synthesized using a lithium catalyst. A material is produced which behaves as though crosslinked at ambient temperatures. This is due to the presence of the two glass transitions associated with each block; the butadiene block has one at 210 K and the styrene block has one at 373 K. Above 373 K plastic flow is observed, but between 210 and 373 K the glassy polystyrene blocks act as crosslinks for the elastomeric polybutadiene and the copolymer exhibits high resilience and low creep characteristics.

The arrangement of the blocks is important; high tensile strength materials, with elastomeric properties similar to a filler reinforced vulcanizate, are obtained only when the copolymer contains two or more polystyrene (S) blocks per molecule. Thus copolymers with the structure {S.B.} or {B.S.B.}, where B is a polybutadiene block, are as brittle as polystyrene, but {S.B.S} and {S.B.S.B} copolymers are much tougher. At ambient temperatures these behave like conventional crosslinked rubbers but they have the additional advantage that their thermal behaviour is reproducible.

The property enhancement of these block copolymers is usually explained in terms of the *"domain concept"*. The glassy polystyrene blocks tend to aggregate in domains (see figure 14.5) which act as both crosslinking points and filler particles. The glassy regions serve to anchor the central elastomeric polydiene blocks securely at both ends and act as effective cross-linking points, thereby precluding the necessity to vulcanize the material.

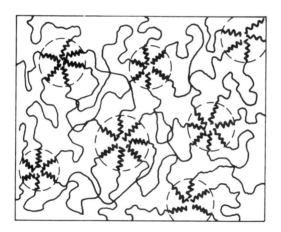

FIGURE 14.5. Schematic representation of elastoplastic sandwich block copolymers, showing areas of aggregation of the glassy {A} blocks, joined by the amorphous rubber-like chains of {B}.

One unexpected application arises from the observation that the presence of more than 10 per cent block copolymer in natural rubber prevents bacterial growth on the polymer surface. Thus, incorporation of the copolymer in butchers' chopping blocks can lead to more hygienic conditions in meat handling.

The synthesis of {ABA} blocks, from a glassy thermoplastic A and an elastomeric B, produces other "elastoplastics" with attractive properties. Polyester chains can be extended with diisocyanate, which is then treated with cumene hydroperoxide to leave a peroxide group at both ends of the chain. By heating this in the presence of styrene, a vinyl polymerization is initiated and an {ABA} block created. The modulus-temperature curves show how the mechanical properties can be modified in this way (figure 14.6).

These block copolymers are known as *thermoplastic elastomers.*

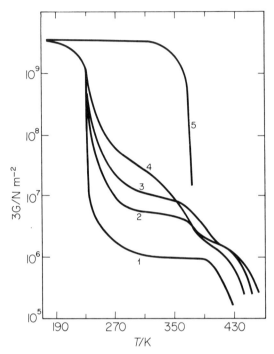

FIGURE 14.6. Modulus-temperature behaviour of polyester-polystyrene block copolymers: 1, polyester; 2, polyester containing 20 per cent of polystyrene; 3, containing 45 per cent polystyrene; 4, containing 60 per cent of polystyrene; 5, pure polystyrene.

14.7 Plasticizers

A polymer sample can be made more pliable by lowering its T_g, and this can be achieved by incorporating quantities of high boiling, low molar mass, compounds in the material. These are called plasticizers and must be compatible with the polymer. The extent to which T_g is depressed depends on the amount of plasticizer present and can be predicted from the relation

$$1/T_g^M = w/T_g + w_l/T_g^l,$$ (14.6)

where T_g^M and T_g^l correspond to the mixture and the liquid, respectively, while w and w_l are the mass fractions of the polymer and plasticizer in the system.

The action of the plasticizer is one of a lubricant, where the small molecules ease the movement of the polymer chains by pushing them further apart. As this lowers both T_g and the modulus, their main use is to increase the flexibility of a polymer for use in tubing and films.

Poly(vinyl chloride), whose T_g is 354 K, usually contains 30 to 40 mass per cent of plasticizers, such as dioctyl or dinonyl phthalate, to increase its toughness and flexibility at ambient temperatures. This depresses T_g to about 270 K and makes the polymer suitable for plastic raincoats, curtains, and "leather-cloth". The low volatility of the plasticizer ensures that it is not lost by evaporation, a mistake made in the early post-war years, which led eventually to a brittle product and considerable customer disaffection. In the rubber industry, plasticizers are usually called oil extenders.

In fibre technology, water absorption is an important factor governing the mechanical response, because the water tends to act as a plasticizer. Thus, as the moisture content increases, the modulus drops, but there is a corresponding improvement in the impact strength. In fibres, such as nylon-6,6, water acts as a plasticizer to depress T_g below room temperature. Thus, when nylon shirts are washed and hung up to drip-dry, the polymer is above the T_g, and this helps creases to straighten out, thereby giving the clothing an "ironed" appearance.

14.8 Crystallinity and mechanical response
The mechanical properties are dependent on both the chemical and physical nature of the polymer and the environment in which it is used. For amorphous polymers, the principles of linear viscoelasticity apply, but these are no longer valid for a semicrystalline polymer.

The mechanical response of a polymer is profoundly influenced by the degree of crystallinity in the sample.

The importance of both crystallinity and molar mass is illustrated by the range of properties displayed by polyethylene. This is shown schematically in figure 14.7 and provides some indication of the effect of these variables.

The interpretation of the mechanical behaviour is further complicated by the presence of glide planes and dislocations, which lead to plastic deformation, but these also serve to provide the materials scientist with a wider variety of property combinations, and can prove useful.

The major effect of the crystallite in a sample is to act as a crosslink in the polymer matrix. This makes the polymer behave as though it was a crosslinked network, but, as the crystallite anchoring points are thermally labile, they disintegrate as the temperature approaches the melting temperature, and the material undergoes a progressive change in structure until beyond T_m, when it is molten. Thus crystallinity has been aptly described by Bawn as a form of "thermoreversible crosslinking".

The restraining influence of the crystallite alters the mechanical behaviour by raising the relaxation time τ and changing the distribution of relaxation and retardation times in the sample. Consequently, there is an effective loss of short τ, causing both the modulus and yield point to increase. The creep behaviour is also curtailed and stress-relaxation takes place over much longer periods. Semi-crystalline polymers are also observed to maintain a relatively higher modulus over a wider temperature range than an amorphous sample.

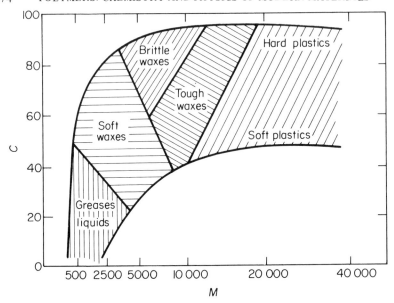

FIGURE 14.7. Influence of crystallinity and chain length on the physical properties of polyethylene. (After Richards, J. *Appl. Chem.*, 1951.) The percentage crystallinity c is plotted against molar mass M.

These points can be illustrated by comparing the elastic relaxation modulus $E_r(t)$ for crystalline (isotactic), amorphous, and chemically crosslinked (atactic), polystyrene samples; see figure 14.8. Crystallinity has little effect below T_g, but as the molecular motion increases above T_g, the modulus of the amorphous polymer drops more sharply. The value of $E_r(t)$ remains high for the crystalline polymer throughout this range until the rapid decrease at the melting temperature is recorded. The crosslinked sample maintains its modulus level at this temperature as the crosslinks are not thermally labile and do not melt.

Rapid quenching of the isotactic polymer destroys the crystallinity and produces behaviour identical to the atactic material. The spherulite size also affects the response; slow cooling from the melt promotes the formation of large spherulites and produces a polymer with a lower impact strength than one cooled rapidly from the melt, whose spherulites are much smaller and more numerous. This effect can be seen as a shift in the damping maxima.

In practical terms, the use of poly(vinyl chloride) in the manufacture of plastic raincoats provides a good illustration of the effect of crystallite crosslinking. The polymer is plasticized, until T_g is below ambient, to make the material flexible, and one might expect that if the coat was hung on a hook (*i.e.* subjected to a tensile load), it would eventually flow onto the floor after prolonged tension. This is not so; the material behaves as though it was a chemically crosslinked elastomer, because it contains a sufficient number of crystallites to act as restraining points and prevent flow.

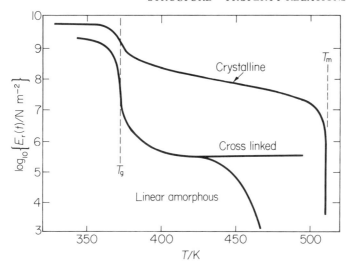

FIGURE 14.8. Illustration of the variation in the modulus–temperature curves for three types of polystyrene.

Similarly, the glass transition of polyethylene is well below ambient temperature, and if the polymer was amorphous, it is likely that it would be a viscous liquid, at room temperature. It is, in fact, a tough, leathery or semi-rigid plastic, because it is highly crystalline and the crystallite crosslinks impart a high modulus and increased strength to the polymer between 188 and 409 K, a very useful temperature range.

The main points can now be restated briefly:

(1) Crystallinity only affects the mechanical response in the temperature range T_g to T_m, and below T_g the effect on the modulus is small.
(2) The modulus of a semi-crystalline polymer is directly proportional to the degree of crystallinity, and remains independent of temperature if the amount of crystalline order remains unchanged.

14.9 Application to fibres, elastomers and plastics
We have seen how various parameters can be altered and combined to produce a material with given responses, but the lines of demarkation dividing polymers into the three major areas of application – fibres, elastomers, and plastics – are by no means well defined. It is most important then to establish criteria which determine that a polymer is a superior fibre, an excellent elastomer, or a particularly suitable plastic before over indulging in the interesting stages of molecular design and engineering.

14.10 Fibres
Superficially a fibre is a polymer with a very high length to diameter ratio (at least 100:1), but most polymers, capable of being melted or dissolved, can be

drawn into filaments. They may, however, have no technical advantages if they cannot meet the requirements of a good fibre; these are high tensile strength, pliability, and resistance to abrasion. In addition, to be useful for clothing, it is preferable that the polymer has $T_m > 470$ K, to allow ironing without damage, but lower than 570 K, to enable spinning from the melt. Also T_g must not be so high that ironing is ineffective. Some typical fibres with useful temperature ranges are shown in table 14.4. These all have T_g lower than 380 K but above room temperature, so that in a cloth the fibres will soften when ironed at about 420 K. This will remove creases or allow pleats to be made which will be retained on cooling. Subsequent washing is normally carried out at temperatures too low to resoften the polymer significantly and so destroy the pleats. This "permanent" crease is a desirable feature of some clothing.

TABLE 14.4. Values of T_m and T_g of some typical fibres

Polymer	Structure	T_g/K	T_m/K
poly(ethylene terephthalate)	$-[(CH_2)_2O \cdot OC - \langle C_6H_4 \rangle - CO \cdot O]-$	343	538
nylon-6,6	$-[NH(CH_2)_6NHCO(CH_2)_4CO]-$	333	538
polyacrylonitrile	$-[CH_2-CH(CN)]-$	378	590
(isotactic) polypropylene	$-[CH_2-CH(CH_3)]-$	268	435

The main distinguishing feature of a fibre is that it is an oriented polymer, and as such is anisotropic, being much stronger along the fibre axis than across it. Thus, the most important technical requirement for fibre formation is the ability to draw or orient the chains in the direction of the fibre axis, and *retain* this after removal of the drawing force. Clearly then, factors which aid this retention of orientation are prime requirements for a good fibre, and these will include all structural features contributing to intermolecular binding.

This means that a polymer should be symmetrical and unbranched to encourage a high degree of crystallinity; it should preferably have a high cohesive energy; and it should have an average length of about 100 nm fully extended. These properties can be conveniently examined under two main headings – the chemical requirements and the mechanical response – and the important factors to consider are: (i) melting and glass transition temperatures; (ii) modulus; (iii) elasticity; (iv) tensile strength, and (v) moisture absorption and dyeability.

CHEMICAL REQUIREMENTS
If the polymer chains are quite short, they are not entangled to any great extent in the solid and are relatively free to move, hence they cannot add to the fibre

strength. As the chain length increases (and so the intertwining), the fibre strength improves and the optimum range of molar mass for a good fibre is 10 000 to 50 000 g mol^{-1}. It has been found that fibre properties deteriorate outside these limits.

We have already mentioned the importance of T_g and T_m and know that these can be affected by chain symmetry, stiffness, and intermolecular bonding. The tensile strength of a fibre is observed to increase with crystallinity, consequently this is a desirable quality and linear chains will be preferred for fibre formation. As the shape and symmetry of a linear chain governs its ability to crystallize, chains containing irregular units, which detract from the linear geometry, should be avoided in fibre forming polymers. This is obvious when comparing terylene I, which is an excellent fibre, with its isomer II, prepared using o-phthalic acid.

$$\sim\!\!\left[(CH_2)_2 \cdot O \cdot OC - \left\langle\bigcirc\right\rangle - CO \cdot O\right]_n\!\!\sim$$

I

$$\sim\!\!\left[(CH_2)_2 \cdot O \cdot OC\diagdown\ \diagup CO \cdot O\right]_n$$

II

Structure II has lost its regularity, is less crystalline, has a lower T_g, and makes a much poorer fibre. This is also true in the polyamide series, where the regular polymer III has $T_m = 643$ K and $T_g = 453$ K

$$\left[NH(CH_2)_6 NH \cdot CO \left\langle\bigcirc\right\rangle CO\right]_n$$

III

$$\left[NH \cdot CH_2 \diagdown\ \diagup CH_2 NH \cdot CO \cdot (CH_2)_4 CO\right]_n$$

IV

but the irregular form has $T_m = 516$ K and $T_g = 363$ K.

Stereoregular polymers also have symmetrical structures and the helices of isotactic polymers can be close packed to produce highly crystalline material. Isotactic polypropylene is crystalline and an important fibre forming polymer, whereas the atactic form has virtually no crystalline content and has little value as a fibre; indeed it is considerably more elastomeric in nature.

Although crystallinity and stereoregularity are important factors in fibre formation, atactic amorphous polymers can also prove useful, if there are inter-molecular forces present. Dipolar interactions between side groups such as $-\hspace{-2pt}(\hspace{-2pt}CN\hspace{-2pt})$ (energy of interaction about 36 kJ mol^{-1}) are significantly stronger than hydrogen bonds or van der Waals forces and serve to improve the molecular alignment immensely. This interaction stabilizes orientation during fibre manufacture and enhances the fibre forming potential of polymers such as polyacrylonitrile and poly(vinyl chloride), both essentially amorphous and atactic. This point highlights the fact that molecular alignment is the most impor-tant factor in fibre formation, not crystallinity, which is only one method of obtaining a stable orientation of chains.

The importance of hydrogen bonding has already been described and will not be dealt with further.

Linear polyesters. Many of the general points discussed can be illustrated conveniently by referring to the numerous linear polyesters which have been prepared. These are grouped together in table 14.5.

A comparison of structure 1 with 2(i) and 2(ii) indicates a drop in T_m caused by the increase in chain flexibility arising from the ethylene and ethylene dioxy groups inserted between the phenylene rings. The change is even more dramatic on comparing 1 with 6 when two phenylene rings are used instead of the $-\hspace{-2pt}(\hspace{-2pt}CH_2\hspace{-2pt})_{\hspace{-2pt}4}$ sequence and the difference in T_m is 205 K.

The influence of symmetry is seen in the terephthalic 3, and isophthalic 4 series. The unsymmetrical ring placement in 4(i) and 4(ii) lowers T_m by 25 K and 77 K respectively, compared with their counterparts 3(i) and 3(ii).

Bulky side groups interfere with the close packing capabilities of a chain, as evidenced by the effect on T_m of the methyl groups in 3(iii), 3(iv), and 5 compared with 3(i) and 3(ii). The additional asymmetry in 3(iv) actually prevents crystallization occurring.

The added stability of secondary bonding in the crystallite is reflected in the increase in T_m in the series 2(i) to 2(iii), as we move from simply van der Waals forces to dipolar and hydrogen-bond interactions. The hydrogen bonding is also sufficient to raise T_m of 2(iii) above that of 3(i), in spite of the extra flexible sequence present in the 2(iii) chain.

These points cover most of the chemical requirements and we can now look at the mechanical properties.

MECHANICAL REQUIREMENTS FOR FIBRES

Fibres are subject to a multitude of mechanical deformations; stretching, abrasion, bending, twisting, shearing, and now the properties of interest are: (i) *tenacity,* which is the stress at the breaking point of the material; (ii) *toughness,* defined as the total energy input to the breaking point; (iii) *initial modulus,* the measure of resistance to stretching (portion A–B of the stress-strain curve, figure 14.9); and (iv) the extent of *permanent set.*

TABLE 14.5. Values of T_m and T_g for linear polyesters

Structure	Group R	T_m/K	T_g/K
1. $-\!\!\left[\,OC-\text{(biphenyl)}-CO\cdot O(CH_2)_2O\,\right]_n$		528	—
2. $-\!\!\left[\,OC-\text{(phenyl)}-R-\text{(phenyl)}-CO\cdot O(CH_2)_2\cdot O\,\right]_n$	(i) $-(CH_2)_4-$ (ii) $-O-(CH_2)_2-O-$ (iii) $-NH-(CH_2)_2-NH-$	443 513 546	— — —
3. $-\!\!\left[\,OC-\text{(phenyl)}-CO\cdot O\cdot R\cdot O\,\right]_n$	(i) $-(CH_2)_2-$ (ii) $-(CH_2)_4-$ (iii) $-CH_2-\underset{CH_3}{\overset{CH_3}{C}}-CH_2-$ (iv) $-CH_2-\underset{CH_3}{CH}-$	538 503 413 non-crystalline	342 353 — 341
4. $-\!\!\left[\,OC-\text{(phenyl, meta)}-CO\cdot O\cdot R\cdot O\,\right]_n$	(i) $\left(CH_2\right)_2$ (ii) $\left(CH_2\right)_4$	513 426	324 —
5. $-\!\!\left[\,OC-\underset{CH_3}{\text{(phenyl)}}-CO\cdot O(CH_2)_2\cdot O\,\right]_n$		343	—
6. $-\!\!\left[\,OC\cdot(CH_2)_4CO\cdot O\cdot(CH_2)_2\cdot O\,\right]_n$		323	—

In technological terminology, the textile industry recognizes the following qualities as suitable: (a) tenacity: 1 to 10 g denier^{-1} (about 5 g denier^{-1} optimum for clothing), (b) modulus of elasticity: 20 to 200 g denier^{-1}, and (c) extensibility: 2 to 50 per cent. The denier is the mass in grams of 9000 m of yarn.

As the mechanical response of a fibre can be controlled to some extent in the spinning process, this will be discussed briefly.

Spinning techniques. The process of converting a bulk polymer sample into a thread or yarn is known as *spinning,* and several methods can be used depending on the nature of the sample.

Melt spinning is used when polymers are readily melted without degradation and the molten polymer is forced through a spinnaret comprising of 50 to 1000 fine holes. On emerging from the holes, the threads solidify, often in an amorphous glassy state, and are wound into a yarn. Orientation and crystallinity are important requirements in fibres and the yarn is subjected to a drawing procedure which orients the chains and strengthens the fibre. This technique is applied to polyesters, polyamides, and polyolefins.

Wet and dry spinning. Acrylic polymers cannot be melt spun because they are thermally labile, and spinning is carried out using concentrated solutions of the polymer. The solvent is removed by evaporation, after extrusion, leaving an amorphous filament, which is then said to have been *dry* spun. When the solution filaments are extruded into a vat of a non-solvent, the polymer precipitates in the form of a thread, and is then a *wet* spun fibre.

Drawing, orientation, and crystallinity. A fibre, in its amorphous state, can be strengthened by *drawing,* a process which extends its length by several times the original, and in doing so aligns the chains in the sample. The process is irreversible and corresponds to the section C-D of the stress-strain curve in figure 14.9, where deformation up to the yield point C is elastic, but beyond this irreversible plastic deformation occurs.

At C the polymer suddenly thins down or "necks" at one point, and subsequent drawing increases the length of the reduced region at the expense of the undrawn region, until the process is complete. Further extension causes rupture at D, the breaking point.

The effect of molecular order is far more important in fibre production than any other area of polymer application and drawing ability is a fundamental requirement in good fibre-forming materials. It is worth pointing out again that crystallinity and orientation are not necessarily synonymous terms and that there is a difference between orientation of crystallites and orientation of chains in the amorphous regions of a polymer. It is the amorphous part of a fibre which will distort and elongate under stress, and these are the areas which must be oriented to improve the intermolecular attraction, if the fibre modulus is to be enhanced. Drawing only improves a highly crystalline fibre to a small extent by orienting the crystallites, but the amorphous fibre is improved immensely.

Drawing affects the mechanical properties of a fibre in several ways. It makes the fibre tough and tenacious, it can increase the modulus and the density, and can alter T_g by orienting the chains in the amorphous regions. For example, the $+O-(CH_2)_2-O+$ group in poly(ethylene terephthalate), which has a *gauche* conformation in the amorphous phase, is "drawn" into the *trans* conformation. This improves the sample crystallinity and T_g also rises 10 to 15 K.

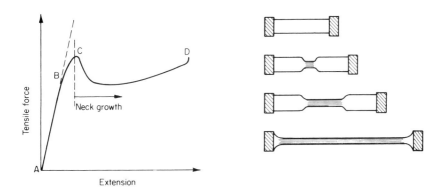

FIGURE 14.9. Successive stages in the drawing of polymer, showing the necking down and subsequent neck growth resulting in increased chain alignment.

The tenacity and physical characteristics of the fibre can also be controlled by the extent of the draw. Limited orientation, produced at low draw stresses, leads to a medium tenacity nylon yarn with low tensile strength, low modulus, and high extensibility, which are all properties associated with a flexible soft material suitable for clothing. Higher draw rates yield high tenacity, high strength yarns, more suited to tyre cord production. Thus some fibre properties are subject to the art of the spinner.

Both tenacity and modulus can be controlled by crystallinity in a fibre. Low pressure polyethylene is highly crystalline and has a fibre tenacity of about 6 g denier^{-1}, but the high pressure, highly branched, and consequently less crystalline polyethylene has a fibre tenacity of only 1.2 g denier^{-1}. We have already seen that conversion of the amide group to a non-hydrogen-bonding group such as a methylol group $-CON(CH_2OH)-$ with formaldehyde, curtails the intermolecular bonding in polyamides. This also increases hydrophilicity and makes the polymer increasingly water soluble, but the hydrophobic characteristics can be restored by methylating the group to $-CON(CH_2OCH_3)-$. At low degrees of substitution the modulus is reduced and a more elastic fibre is obtained. As the substitution increases, the crystallinity is completely destroyed, and the fibre forming capacity disappears.

Modulus and chain stiffness. Fibre modulus can be regulated by orientation and crystallinity, but a third parameter, chain stiffness, is available for modification, if additional control is required.

The effect of chain stiffness on the initial modulus is seen in figure 14.10.

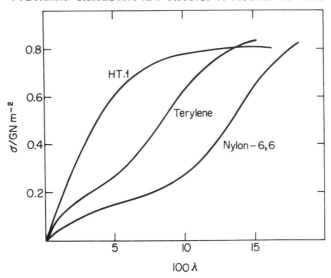

FIGURE 14.10. Stress-strain curves for three fibres, showing the changes wrought by the introduction of one (terylene), then two (HT.1) phenylene groups into the repeat unit of the chain, the stress σ is plotted against 100λ the percentage elongation.

The increase in chain rigidity on moving from nylon-6,6 $\text{+NH(CH}_2)_6\text{NHCO}$ $(\text{CH}_2)_4\text{CO+}_n$ to terylene $\left[\text{O(CH}_2)_2\text{O.C}-\bigcirc-\overset{\text{C}}{\underset{\text{O}}{\parallel}}\right]_n$ to poly(m-phenylene isophthalimide), HT.1, $\left[\text{NH}-\bigcirc-\text{NH}-\overset{\text{C}}{\underset{\text{O}}{\parallel}}-\bigcirc-\overset{\text{C}}{\underset{\text{O}}{\parallel}}\right]_n$

is manifest in an increase in the initial modulus, with the inclusion of aromatic rings in the chain. The advantages can be lost if carried to extremes, as with poly(p-phenylene), where the chain of catenated rings is now excessively rigid, intractable, and unsuitable for fibre formation.

Other factors. The moisture regain of a fibre is important when comfort is being considered. In hot weather the ability to absorb perspiration makes clothing more comfortable, and polar polymers are best adapted for this purpose. High moisture retention also decreases the resistivity of the fibre and reduces the tendency to build up static charges which attract dirt and increase the discomfort. Most synthetic fibres have poor moisture regain characteristics and have to be modified in some way to improve this defect. Grafting of poly(ethylene oxide) or acrylic acid on to nylons improves the moisture uptake immensely without affecting the mechanical properties.

Dyeing is also a problem and chemical modification is often necessary. Sites are provided for dyeing by substituting a number of $-SO_3H$ groups in the phenylene rings of the terylene chain or by copolymerizing acrylonitrile with small quantities of vinyl sulphonic acid. These modifications also improve the moisture uptake.

When selecting a fibre for clothing, one should avoid material with a high value of "permanent set". This is a measure of the amount of irreversible flow (C-D in figure 14.9) left in the polymer and is reflected in an increase in fibre length after being subjected to a stress. Obviously in clothing, where the amount of knee or elbow bending is great, a high permanent set value will result in gross fibre deformation and "baggy trousers" or "kneed" stockings. This is partially or totally offset by drawing, but high draw ratios may make the fibre hard. Hence, while poor creep recovery in an article results in loss of shape, its capacity to absorb energy may deteriorate, and if overcompensated for, may result in actual material failure.

It is not uncommon to be forced into a compromise when faced with choice between two incompatible properties.

14.11 Elastomers and crosslinked networks

Rubber-like elasticity and its associated properties have already been discussed in some detail (see section 11.7 and chapter 13) and only a brief resumé of the relevant features will be given.

The fundamental requirements of any potential elastomer are that the polymer is amorphous with a low cohesive energy, and that it is used at temperatures above its glass transition. The polymer in the elastic region is characterized by a low modulus (about 10^5 N m^{-2}) and, for useful elastomers, by large reversible extensions. This reversibility of the slippage of flow units requires a chain in which there is a high localized mobility of segments, but a low overall movement of chains relative to one another. The first requirement is satisfied by flexible chains, with a low cohesive energy, which are not inclined to crystallize (although the development of some crystalline order on stretching is advantageous). The second requirement, prevention of chain slippage, is overcome by crosslinking the chains to form a three-dimensional network.

Crosslinking. Crosslinking provides anchoring points for the chains and these anchor points restrain excessive movement and maintain the position of the chain in the network. This is not confined to elastomers, however, and the improved material qualities which result are also found in the crosslinked phenolformaldehyde, melamine, and epoxy resins.

When a sample is crosslinked, (1) the dimensional stability is improved, (2) the creep rate is lowered, (3) the resistance to solvents increases, and (4) it becomes less prone to heat distortion, because T_g is raised. All these effects tend to be intensified as the crosslink density is increased and can be controlled by adjusting the number of crosslinks in a sample.

Creep in crosslinked polymers. The creep response depends mainly on the temperature and the crosslink density. At temperatures below T_g, crosslinking has little effect on the properties of the material, but above T_g, secondary creep, arising from irreversible viscous flow, is reduced or eliminated by crosslinking.

Creep is a function of the elastic modulus, the mechanical damping, and the difference between ambient temperature and T_g. The thermosetting resins usually have a high modulus, low damping characteristics, and T_g well above ambient, consequently the creep rate is low and they have good dimensional stability.

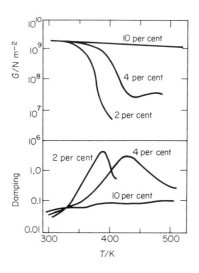

FIGURE 14.11. Influence of crosslinking on the dynamic mechanical response of a phenol-formaldehyde resin. Concentrations of the crosslinking agent hexamethylenetetramine are shown alongside the appropriate curve. (After Nielsen.)

The effect of increasing the crosslink density on these parameters, is illustrated for a phenol-formaldehyde resin in figure 14.11.

Above T_g the modulus is a function of the extent of crosslinking; the damping peaks shift to higher temperatures as T_g increases and eventually become difficult to detect.

This shows the extent to which crosslink density can affect the physical behaviour.

Additives. Many elastomers are subject to oxidative degradation and can be protected to some extent by the addition of antioxidants, such as amines and hydroquinones.

The abrasion resistance can also be improved by adding a filler to reinforce the elastomer, and carbon black is widely used for this purpose. Fillers (glass fibre, mica, sawdust) are also used in the thermosetting resins as reinforcement.

14.12 Plastics

So far attention has been focused predominantly on fibre and elastomer requirements, because it is considerably more difficult to be specific about the qualities desired in a plastic material when the range of applications covered is very much more extensive. The general principles relating to the control of T_m, T_g, modulus, *etc.*, can all be applied to the formation of a specific type of plastic and we shall simply try to illustrate briefly the diversity of problems encountered in the field of plastic utilization.

The conflict between low creep and high impact strength mentioned earlier is not confined to fibres, but is also a problem encountered in plastic selection. It is an important point to consider for the engineering requirements of the material, when the ability to absorb energy is desirable, but is at odds with the equally desirable qualities of high rigidity and low creep. The problem to be faced is then how to make a brittle, glassy polymer tougher, *i.e.* how to limit the modulus or tensile strength. In general, an increase in crystallinity (and consequently the modulus) tends to make a plastic more brittle. Crystallinity can be controlled by copolymerization or branching and the brittleness can be tempered using one or other of these modifications. Alternatively, an elastomeric component can be introduced, which will improve the impact strength by reducing the rigidity and yield stress. This has been used in "high impact" polystyrene or acrylonitrile-butadiene-styrene (ABS) copolymers, where the elastomeric component is above its T_g under prevailing environmental conditions and acts as a second phase. This leads to an increased damping efficiency which is manifest in the appearance of a second low temperature damping maximum in the damping curve. This is seen in figure 14.12 for "high impact" polystyrene-butadiene copolymer (SBR rubber) whose T_g is 213 K. The phenomenon is similar to the toughening effect in semi-crystalline polymers caused by the strengthening of the amorphous regions with crystalline crosslinks, but in the latter case, the two phase aspect arises from the existence of crystalline and amorphous regions.

While orientation is most important in fibre formation it can also improve the response of a brittle polymer and increase its ductility. This is particularly true in film preparation or moulding where viscous flow is inclined to introduce a certain degree of chain alignment at some stage in the process.

Interchain interactions also affect performance and poly(oxymethylene) has a higher modulus in the glassy state than polyethylene presumably because of the polar attractions between the chains.

When faced with the problem of selecting a plastic for a given purpose, a design engineer must then be concerned with the properties of the material, the ease of processing or fabrication, the behaviour under the environmental conditions the product will be subjected to (*i.e.* the thermal range), and, of course, the economic factors. Each problem has to be treated as a specific case and familiarity with structure-property relations aids the selection. The illustrations are limited to two widely differing aspects.

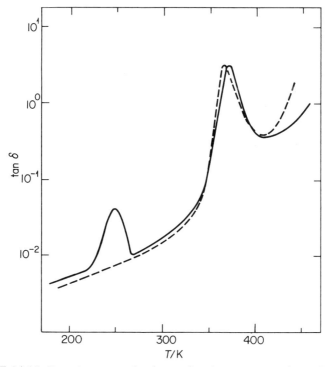

FIGURE 14.12. Damping curves for (– ·· –) polystyrene, and (———) high impact polystyrene. The latter has an internal friction peak below ambient. The damping tan δ is plotted against temperature T.

Plastic selection for bottle crate manufacture. The difficulties encountered when choosing a suitable plastic for a particular use arise mainly because each case is associated with a unique combination of properties. A good example, concerning bottle crate manufacture, has been cited by Willbourn.

High density polyethylene was chosen for the manufacture of beer crates in West Germany, because it was the cheapest plastic available which was sufficiently tough and rigid for the purpose. It was also found to have a satisfactory creep response and good impact resistance down to 253 K, which is adequate for continental winter temperatures. This plastic and the crate design were suitable for the use pattern in West Germany, where crates were usually piled 12 high.

When these crates were used in the U.K., where the practice is to stockpile 20 to 36 high for much longer periods, a rapid rate of crate failure was experienced. The change in conditions necessitated a new choice of plastic. This had to have better creep properties and a higher rigidity, but did not have to retain those good qualities at temperatures below 263 K, because of the milder U.K. winters. Poly(vinyl chloride) was considered but is too difficult to mould;

polystyrene and polypropylene have good creep characteristics but these deteriorate at lower temperatures. The problem was solved by using poly(propylene-*b*-ethylene) copolymers, which have a good toughness and mechanical response in the required temperature range. This is shown in figure 14.13 where the high density polyethylene failed under a load of 1000 kg in 29 h, but the copolymer survived two months.

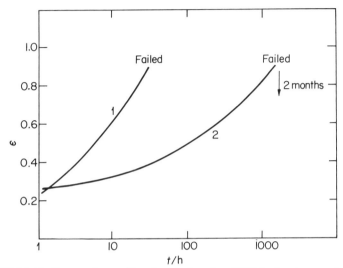

FIGURE 14.13. Comparison of loading tests using 1000 kg load on crates made from (1) high density polyethylene and (2) poly(ethylene *b*-propylene) copolymer. (After Wilbourn, *Plastics and Polymers*, 1969.) The percentage compressive strain ϵ is plotted against time t.

In this instance, both environmental conditions and industrial practice were important factors.

Medical applications. The use of polymeric materials in the medical field is growing and raises problems peculiar to the mode of application. Prosthesis is one of the major medical interests and certain plastic replacement parts are now commonly used. High density polyethylene is a successful replacement part for damaged hip joints and is employed as the socket, which accommodates a steel ball cemented to the femur using poly(methyl methacrylate). Artificial corneas can be prepared from poly(methyl methacrylate), while sections of artery are replaced by woven nylon or terylene tubes. Heart valves have been made from polycarbonates and even artificial hearts, made from silicone rubber, have met with limited success. Plastic replacements for nose and ear cartilage; body absorbing sutures; and the use of polymeric membranes for dialysis in artificial kidney machines are only a few examples in a steadily growing list of uses.

The selection of suitable polymers for medical use focuses attention on the inertness of the polymer, its mechanical properties, and the extent of its biostability. It is useless implanting a polymer in the body which will be rejected or will degrade to produce toxic materials. The sample should also be pure and free of plasticizer, which might leach out and cause harmful side effects. The polymer has to be resistant to mechanical degradation and particularly abrasion, in case the abraded particles act as irritants. These conditions tend to limit the choice.

The use of polymers as adhesives is of particular interest. One reason for using the α-cyano acrylate esters as tissue adhesives has been the observation that there is a progressive, non-toxic, absorption of the substances by the body. Interest in the use of polymers as reagents, which actively take part in the body functions, is now being evaluated.

External to the body, "hydrogels" are used as contact lenses and their use may be extended to implantation. They are composed of crosslinked networks of hydroxyl methacrylate copolymers which swell when in contact with water.

Films and membranes are also used. A patient can be encased in a poly(vinyl chloride) tent while germ-free air is pumped through the canopy. It has also been suggested that films which allow the selective passage of oxygen in one direction could be used as oxygen tents. These would be similar to the silicone membranes which allow predominant passage of oxygen from water to air, and have been used to make a cage capable of supporting underwater, non-aquatic life in an air atmosphere extracted from the water.

This expanding field will no doubt demand new polymers with specific applications.

14.13 High temperature fibres and plastics

The search for new thermally-stable fibres and plastics has accelerated in the past decade, and has met with considerable success. While thermally stable polymers can by synthesized, fabrication of these often intractable materials may be extremely difficult and presents a major problem. For example, polyindigo can be prepared by oxidative coupling, and while the polymer is apparently stable up to at least 870 K, no solvent can be found to dissolve it for fabrication, neither will it form a stable melt.

The main approach to the synthetic problem has been to introduce cyclic structures into the chain to stiffen it and both phenylene and heterocyclic groups are used separately or in combination. The aromatic polyamides are good examples of fibres with good thermal stability, and other structures such as

poly(p-phenylene) , poly(2,2'-(m-phenylene)-5,5'-bibenzimidazole), (PBI) and the ladder polymer poly(benzimidazobenzophenthroline) all show marked resistance to thermal breakdown.

The search for condensation polymers with good thermal stability has been aided by the development of low temperature techniques to enable the temperature sensitive units to be incorporated in the chain, and at the same time allowing fibres to be spun directly from the interface. This aspect is dealt with in chapter 2, where other thermally stable polymers are also described.

A considerable amount of molecular engineering has taken place in this field, especially in the aromatic polyamide series. Here a balance has been sought, and as many phenylene units as possible are incorporated in the chain, while making use of flexible $-\!\!\!+\!O\!+\!\!\!-$ units to maintain solubility.

While an in-depth exploration of this field cannot be undertaken here, it is worth examining one important product, the carbon fibre.

14.14 Carbon fibres

Although originally studied for its high temperature qualities, the carbon fibre is, at present, used to advantage mainly in low temperature situations.

The fibres are prepared by converting oriented acrylic fibres into aligned graphite crystal fibres in a two-stage process. In the first stage the acrylic fibre is oxidized, under tension to prevent disorientation of the chains, by heating in a current of air at 490 K for several hours. This is thought to lead to cyclization and the formation of a ladder polymer.

idealized structure

The second stage involves heating the fibres for a further period at 1770 K to eliminate all elements other than carbon. This *carbonization* is believed to involve crosslinking of the chains to form the hexagonal graphite structure, and this final heat treatment can affect the mechanical properties to a marked extent as shown in figure 14.14. The major application, so far, is in composite structures where they act as extremely effective reinforcing fibres. These reinforced plastic composites find uses in the aircraft industry, in the small boat trade, and as ablative composites.

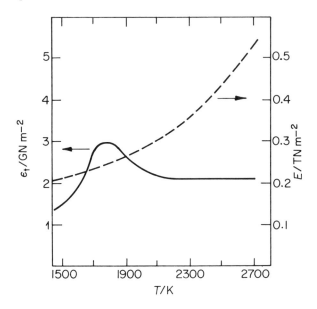

FIGURE 14.14. Mechanical properties, the tensile strength ϵ_t and Young's modulus E, of carbon fibres as a function of the graphitizing temperature I. (After Bailey and Clarke, *Chem. in Brit.*, 1970.)

14.15 Concluding remarks

The systematic study of structure-property relations provides an understanding of many of the fundamentals of the subject and can lead to quick dividends, as shown by the following example of the advanced art of fibre engineering.

A sheep grows a wool fibre which possesses a corkscrew crimp in the dry state, and affords the animal a bulky insulation blanket. When it rains the fibre becomes wet and loses the crimp; the wool strands then bed down to form a close packed, rain-tight, covering capable of reducing body heat loss in the damp conditions. This *evolutionary* wool fibre, whose properties are derived from its bicomponent nature, was simulated in the laboratory by preparing a two component acrylic fibre in which each component had a different hydrophilicity. This feat was achieved by fibre scientists in only a few months by making use of their knowledge of structure and behaviour.

Of course, nature was first, and, at best, only reasonable facsimiles of some natural products can be synthesized in the laboratory. The scientist is not yet able to match the sophistication of many naturally occurring macromolecules, which are no longer simply materials, but working functional units. The complexity of the inter-relation between structure and function in many proteins and nucleo-proteins is the ultimate in molecular design, and while it will be some time before we can hope to reach this level in synthesis, progress in understanding the relations in simpler systems is a step in the right direction.

General Reading

C. E. H. Bawn, "Structure and performances", *Plastics and Polymers,* 373 (1969).
F. W. Billmeyer, *Textbook of Polymer Science.* John Wiley and Sons (1962).
B. Bloch and G. W. Hastings, *Plastics in Surgery.* Thomas (1967).
J. A. Brydson, *Plastic Materials.* Iliffe Books Ltd. (1966).
W. Bruce-Black, "Structure-property relationships in high temperature fibres",
    Trans. N. Y. Acad. Sci., 32, 765 (1970).
I. Goodman, *Synthetic Fibre Forming Polymers.* R.I.C. (1967).
J. W. S. Hearle and R. H. Peters, *Fibre Structure.* Butterworths (1963).
R. W. Moncrieff, *Man-made Fibres.* John Wiley and Sons (1963).
F. Rodriguez, *Principles of Polymer Systems.* McGraw-Hill (1970).
P. D. Ritchie, *Plasticizers, Stabilizers and Fillers.*
R. B. Seymour, *Introduction to Polymer Chemistry,* Chapter 14. McGraw-Hill
    (1971).
E. A. Tippetts and J. Zimmerman, "Polymers as fibres", *J. Appl. Pol. Sci.,*
    8, 2465 (1964).
A. V. Tobolsky and H. Mark, *Polymer Science and Materials,* Chapters 14 and 15.
    Wiley-Interscience (1971).

References

1. J. E. Bailey and A. J. Clarke, *Chem. in Britain,* 6, 484 (1970).
2. M. F. Drumm, C. W. H. Dodge, and L. E. Nielsen, *Ind. Eng. Chem.,* 48, 76
    (1956).
3. R. B. Richards, *J. Appl. Chem.,* 1, 370 (1951).
4. A. H. Willbourn, *Plastics and Polymers,* 417 (1969).

Index